Number 4 · 1981 · Volume 1

Acta Biotechnologica

Journal of microbial, biochemical and bioanalogous technology

Akademie-Verlag
Berlin

Acta Biotechnologica **1** (1981) 4, 309–392
31 007 EVP 30,– M ISSN 0138-4988

Instructions to Authors

1. Only original papers that have not previously been published will be accepted.
Manuscripts may be submitted in English, German or Russian but always in duplicate. The name of the institute (inclusive full address) from which the manuscript originates should be stated below the author's name(s). The authors are responsible for the contents of their contributions.

2. Original papers should not consist of more than 20 typewritten pages (2 lined-spacing), including references, tables and figures; short original communications may have a maximum of 6 typewritten pages.

3. Each paper should be preceded by two summaries. One summary in the language of the paper and one in another language. In each case one summary must be in english.

4. Latin names of species as well as passages to be printed in italics for greater emphasis should be underlined. Please use only units and symbols of the SI-system.

5. Tables may be used to effect a reduction of the text or to make assimilation easier. They should be numbered consecutively throughout and supplied with a brief heading. They shouldn't be written among the text but on separate sheets.

6. Number and size of illustrations should be limited to the essential minimum, they should be numbered consecutively and brought on separate sheets.
Line drawings including graphs and diagrams should be drawn in black ink. Half-tone illustrations should be supplied as white glossy prints. Figure legends are to be typed in sequence on a separate sheet.
Authors name is to be written on to the rear of each sheet.

7. References, listed at the end of the contribution, should contain only works cited in the text. They should be numbered in the order in which they are at first mentioned in the text. Please give surnames and initials of all authors, name of journal abbreviated according to "Chemical Abstracts — List of Periodicals", volume number, year of publication, issue number or month, first page number; books are to be cited with full title, edition, volume number, page number, place of publication, publisher and year of publication.

8. Notes on the text may be given as footnotes on the relevant page.

9. 50 reprints are free of charge.
Additional reprints may be ordered.

10. The author receives 2 galley proofs for correction.
They are to be sent back to the redaction
(Dr. Dimter, DDR-7050 Leipzig, Permoserstr. 15)
as soon as possible.

ISSN 0138-4988

Acta Biotechnologica

Journal of microbial, biochemical and bioanalogous technology

Edited at
Institute of Technical Chemistry of the Academy of Sciences of the G.D.R.; Leipzig
and
Institute of Technical Microbiology; Berlin
by
M. Ringpfeil, Leipzig and G. Vetterlein, Berlin

Number 4

1981

Volume 1

AKADEMIE-VERLAG · BERLIN

"Acta Biotechnologica" publishes reviews, original papers, short communications and reports out of the whole area of biotechnology.
The journal shall promote the foundation of biotechnology as a new, homogeneous scientific field.
According to biotechnology are microbial technology, biochemical technology and technology of synthesyzing and applying of bioanalogous reaction systems.
The technological character of the journal is guarenteed thereby that microbial, biochemical, chemical and physical contributions must show definitely the technological relation.

Terms of subscription for the journal "Acta Biotechnologica"
Orders can be sent

- **in the GDR:** to Postzeitungsvertrieb, to a book-shop, or to Akademie-Verlag, DDR - 1080 Berlin, Leipziger Straße 3–4;
- **in the other socialist countries:** to a book-shop for foreign languages literature or to the competent news-distributing agency;
- **in the FRG and Berlin (West):** to a book-shop or to the wholesale distributing agency Kunst und Wissen, Erich Bieber, D - 7000 Stuttgart 1, Wilhelmstr. 4–6;
- **in the other Western European countries:** to Kunst und Wissen, Erich Bieber GmbH, CH - 8008 Zürich, Dufourstraße 51;
- **in other countries:** to the international book- and journal-selling trade, to Buchexport, Volkseigener Außenhandelsbetrieb der DDR, DDR - 7010 Leipzig, Postfach 160, or to Akademie-Verlag, DDR - 1080 Berlin, Leipziger Straße 3–4.

Acta Biotechnologica

Herausgeber: Institut für technische Chemie der AdW,
DDR - 7050 Leipzig, Permoserstr. 15
(Direktor: Prof. Dr. Manfred Ringpfeil) und
Institut für Technische Mikrobiologie
DDR - 1017 Berlin; Alt-Stralau 62
(Direktor: Dipl. Ing. G. Vetterlein)

Verlag: Akademie-Verlag, DDR - 1080 Berlin, Leipziger Straße 3–4;
Fernruf: 2236221 und 2236229; Telex-Nr.: 114420; Bank: Staatsbank der DDR, Berlin, Konto-Nr.: 6836-26-20712.

Redaktion: Dr. Lothar Dimter (Chefredakteur),
Martina Bechstedt (Redakteur),
DDR - 7050 Leipzig, Permoserstr. 15; Tel.: 6861255.

Veröffentlicht unter der Lizenznummer 1671 des Presseamtes beim Vorsitzenden des Ministerrates der Deutschen Demokratischen Republik.

Gesamtherstellung: VEB Druckhaus „Maxim Gorki", DDR - 7400 Altenburg.

Erscheinungsweise: Die Zeitschrift „Acta Biotechnologica" erscheint jährlich in einem Band mit 4 Heften. Bezugspreis eines Bandes 120,– M zuzüglich Versandspesen; Preis je Heft 30,– M.

Bestellnummer dieses Heftes: 1094/1/4.

AN (EDV) 42133

Acta Biotechnologica 1 (1981) 4, 311–325

Verfahrenstechnische Grundlagen der Reaktorgestaltung

M. ZLOKARNIK

Bayer AG, D-5090 Leverkusen, BRD

Vortrag gehalten anläßlich des 2. Symposiums der sozialistischen Länder über Biotechnologie, Leipzig 2.–5. 12. 1980

Summary

Starting with the definition of the process term k_La, steady state and nonsteady state measuring methods are described for its determination. Then the sorption characteristics for mixing vessels and for bubble columns are presented with respect to the coalescence behaviour of the system treated. They permit the scale-up of these devices and the optimization of their process parameters for a required oxygen uptake. In addition to the sorption characteristics for the given system the knowledge of the flooding point and the power characteristics is necessary for the lay-out of mixing vessels, whereas in the case of bubble columns the gas hold-up characteristic needs to be known.

Zusammenfassung

Ausgehend von der Definition der verfahrenstechnischen Zielgröße k_La werden Methoden zu ihrer Bestimmung beschrieben, wobei zwischen instationären und stationären Meßmethoden unterschieden wird.
Unter Berücksichtigung der Problematik der Gasblasenkoaleszenz und ihres Einflusses auf den k_La-Wert werden Sorptionscharakteristiken für verschiedene Rührbehälter- und Blasensäulensysteme abgeleitet. Diese Sorptionscharakteristiken erlauben die Dimensionierung und Optimierung von Rührbehältern und Blasensäulen bezüglich des geforderten Sauerstoffeintrages. Während zur Auslegung von Rührbehältern weiterhin die Kenntnis der sogenannten Überflutungs- und Leistungscharakteristiken im Stoffsystem gasförmig/flüssig erforderlich sind, wird im Falle der Blasensäule die Kenntnis des Gasvolumens in der Flüssigkeit ("gas hold up") benötigt.

Vorbemerkung

Die Bioreaktoren für aerobe mikrobiologische Prozesse werden primär nach Maßgabe des geforderten O_2-Eintrages konzipiert und dimensioniert; die übrigen verfahrenstechnischen Fragestellungen wie Wärmetransport, Rückvermischung des Reaktorinhaltes und die Scherbeanspruchung der Biomasse sind meist von sekundärer Bedeutung.

Definition der Zielgröße $k_L a$

Die Zielgröße des Stofftransportes im System gasförmig/flüssig ist der sogenannte „Stofftransportkoeffizient" $k_L a$, welcher mit der allgemeinen Stofftransport-Beziehung definiert ist:

$$G/V = k_L a \Delta c; \quad k_L a = G/(V \Delta c) \tag{1}$$

Es bedeuten

G/V — volumenbezogener Massenstrom durch die Phasengrenze
k_L — flüssigkeitsseitiger Stoffübergangskoeffizient
a — volumenbezogene Phasengrenzfläche
Δc — charakteristische Konzentrationsdifferenz

Beziehung (1) impliziert Volumenbegasung beim Vorliegen einer quasiuniformen Gas/Flüssigkeits-Dispersion (Volumenbezogenheit) sowie Vernachlässigbarkeit des gasseitigen Stoffübergangswiderstandes ($k_G \gg k_L$). Bei der Frage, von welchen Einflußgrößen $k_L a$ abhängen kann, wird man davon ausgehen, daß $k_L a$ eine volumenbezogene, also intensiv formulierte Größe ist, die daher per definitionem von den geometrischen Parametern der Vorrichtung nicht abhängt. Sie ist auch von den Stoffparametern der Gasphase unabhängig, da $k_G \gg k_L$ vorausgesetzt wurde.

Bestimmung von Δc

Das sogenannte „treibende Konzentrationsgefälle" Δc kann nur bei einer Oberflächenbelüftung durch

$$\Delta c = c_S - c \tag{2}$$

gegeben sein (c_S — Sättigungskonzentration des Gases in der Flüssigkeit bei gegebener Temperatur und gegebenem Druck; c — augenblickliche Konzentration des Gases in der Flüssigkeit), weil nur hier c_S zeitlich und örtlich konstant bleibt. Bei der Volumenbelüftung wird sich c_S sowohl örtlich (wegen unterschiedlicher hydrostatischer Drücke) al- auch zeitlich (Änderung des Molenbruchs x des zu absorbierenden Gases in aufsteigendes Luftblase infolge Verarmung) ändern. Hier läßt sich bei stationärer Arbeitsweise dar charakteristische Konzentrationsgefälle mit einer logarithmischen Konzentrations differenz beschreiben, die auf Annahmen beruht: a) vollständige Rückvermischung der Flüssigkeit im System und damit $c = \text{const}$, b) vollständige Segregation der Gasphase (zutreffend bei nicht-koaleszierenden Systemen und bei koaleszierenden Systemen bei Begasungsdichten $\leqq 300$ eff. m^3/m^2h). Hierfür gilt

$$\Delta c_m = (c' - c'')/\ln[(c' - c)/(c'' - c)], \tag{3}$$

wobei c' und c'' die Sättigungskonzentrationen am Gaseintritt und Gasaustritt bedeuten:

$$c' = c_S(\vartheta)\, x'(p + \varrho g H),$$
$$c'' = c_S(\vartheta)\, x'' p.$$

p ist der Gasdruck über der Flüssigkeit, x' und x'' die Molenbrüche des zu absorbierenden Gases. x'' im Abgas kann man berechnen, wenn der O_2-Eintrag G, der Gasdurchsatz q sowie die O_2-Dichte ϱ_{O_2} bekannt sind:

$$x'' = (q x' - G/\varrho_{O_2})/(q - G/\varrho_{O_2}).$$

Da k_La mittels Δc berechnet wird (vgl. Bez. (1)) und somit sein Zahlenwert von dem verwendeten Δc-Ausdruck abhängt, ist es erforderlich, bei der Angabe von k_La-Werten stets auch die zugrunde liegende Definition von Δc anzugeben; vgl. hierzu besonders [9]. In diesem Aufsatz werden k_La-Werte ausschließlich mit Δc gemäß (3) gebildet.

Methoden zur k_La-Bestimmung

Je nachdem, ob sich die Konzentration c des in der Flüssigkeit gelösten Gases während des Vorganges zeitlich ändert oder konstant bleibt, unterscheidet man stationäre und instationäre Versuchsbedingungen:

Instationäre Meßmethoden:

a) Verfolgung von $c(t)$ beim Aufsättigen (oder bei der Desorption) mittels O_2-Elektroden (beschränkt auf Verwendung von O_2 bzw. Luft und von Wasser bzw. wäßrigen Lösungen). b) Manometrische Methode [25] (beliebige Flüssigkeit und beliebiges, jedoch reinstes Gas).

Stationäre Meßmethoden:

a) Hydrazin-Methode [25] ($N_2H_4 + O_2 \rightarrow H_2O + N_2$) gestattet Arbeiten bei Salzkonzentrationen < 1 g/l, wodurch sogenannte koaleszierende Bedingungen wie beim reinen Wasser vorliegen; hierbei muß allerdings $c \gg 0 =$ const. mit einer O_2-Elektrode verfolgt werden. b) Sulfit-Oxidation ($Na_2SO_3 + 1/2\,O_2 \rightarrow Na_2SO_4$) verläuft so schnell, daß $c = 0$ ist, dafür liegt hier mit einer Salzkonzentration von $\geqq 10$ g/l ein nicht-koaleszierendes Stoffsystem vor (vgl. nächstes Kapitel).
Der O_2-Eintrag in die Fermentationslösung hängt von stofflichen und prozeßbedingten Parametern ab. Bei wasserdünnen Lösungen ist die Kenntnis ihres Koaleszenzverhaltens, bei viskoseren Stoffsystemen die ihres Viskositätsverhaltens (nicht-Newtonsches, elastoviskoses usw.) sowie ihrer effektiven Viskosität zur Auslegung der Begasungsvorrichtung unabdingbar.

Blasenkoaleszenz und ihr Einfluß auf k_La

In reiner Flüssigkeit dispergierte feine Gasblasen haben die Tendenz, zu größeren zusammenzulaufen. Dieser, als Koaleszenz von Gasblasen bezeichnete Vorgang kommt dadurch zustande, daß der Flüssigkeitsfilm (Flüssigkeitslamelle) zwischen zwei benachbarten Gasblasen ausdünnt, bis er platzt ("thinning", "drainage"). Dieser Vorgang ist extrem schnell bei reinen Flüssigkeiten und extrem langsam bei zum Schäumen neigenden Lösungen: Die Fähigkeit zum Schäumen spiegelt direkt das Koaleszenzverhalten des Systems wider.
Der Grund für das Ausdünnen des Flüssigkeitsfilms besteht in den Druckdifferenzen zwischen dem Film und dem Kern der Flüssigkeit. Diese werden durch unterschiedliche Kapillardrücke hervorgerufen, die von den unterschiedlichen Krümmungsradien und von den van der Waalsschen Kräften herrühren. Diese Druckdifferenzen Δp stehen jeweils mit der Differenz in der Oberflächenspannung σ im Kräftegleichgewicht:

$$\delta \Delta p = 2 \Delta \sigma \cdot \delta - \text{Filmdicke}$$

Wir müssen uns nun daran erinnern, daß im Falle von Lösungen eine Vergrößerung der Oberfläche zu einer Änderung der Oberflächenkonzentration führt, was wiederum die Oberflächenspannung verändert. Aus dem Gibbsschen Satz über die Veränderlichkeit der Oberflächenspannung einer Lösung mit deren Konzentration wissen wir, daß sich die Oberflächenkonzentration Γ_i des betrachteten Gelösten i bei der Vergrößerung der Oberfläche o in einer ganz bestimmten Weise ändert:

$$\Delta\Gamma_i = \frac{dc_i}{do} = -\frac{c_i}{RT}\frac{d\sigma}{dc_i}$$

Bei positivem Ausdruck $d\sigma/dc$ tritt auf der Oberfläche eine Verarmung, im umgekehrten Fall eine Anreicherung der gelösten Substanz ein. Nach der Koaleszenztheorie von MARUCCI [11] ist jedoch dessen ungeachtet die im Verlaufe des individuellen Koaleszenzvorganges beträchtliche Oberflächenspannung (Nach [11] dünnt die Lamelle zu einigen Hundert Å aus, bevor sie platzt, wobei der Ausdruck $\Delta o/V$ Werte bis zu 10^6 cm^{-1} annehmen kann) immer mit einem starken Anstieg der Oberflächenspannung verbunden. Die Größe $d\sigma/dc$ übt daher einen großen Einfluß auf die Dynamik des Films aus und es ist zu erwarten, daß sie sich zur Voraussage des Koaleszenzverhaltens von Lösungen

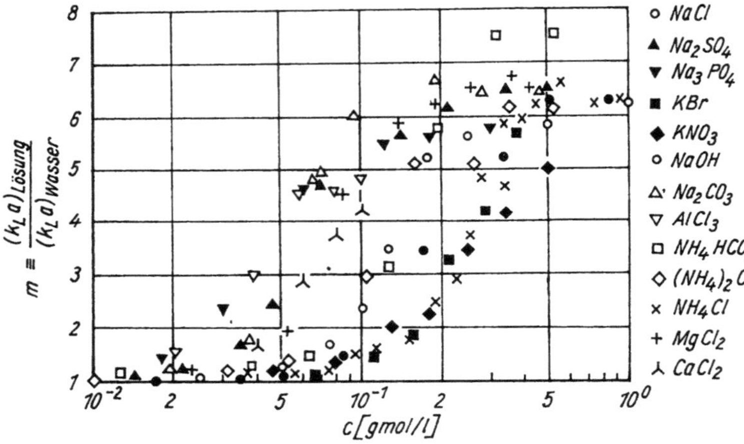

Abb. 1. Beschleunigungsfaktor m als Funktion der Salzart und Salzkonzentration. Aus [30].

eignet. (Bei reinen Flüssigkeiten ist dagegen $d\sigma/dc = 0$; daher kommt es in der ausdünnenden Lamelle zu keinem Anstieg der Oberflächenspannung, was den Koaleszenzvorgang verlangsamen bzw. unterbinden könnte)

Begast man dünnflüssige reine Flüssigkeiten wie Wasser, Methanol oder Aceton, so resultiert unabhängig von der Gaszerteilung ein stabiler Blasendurchmesser von 3—5 mm, sobald die feinsten, primär gebildeten Gasbläschen den Bereich hoher Scherkräfte verlassen haben. Eine Erzeugung von feinsten Primär-Gasbläschen in reinen Flüssigkeiten ist daher unökonomisch.

In Lösungen von Salzen oder ineinander löslichen Flüssigkeiten wird dagegen die Koaleszenz von Primär-Gasbläschen stark unterdrückt; die primär erzeugte Blasengröße wird desto besser konserviert, je höher die Konzentration der Lösung ist. Die resultierende Gasblasengröße beträgt hier 0,2—0,5 mm und ist somit um eine ganze Größenordnung kleiner als bei reinen Flüssigkeiten.

Um quantitative Aussagen über den Einfluß von anorganischen Salzen (Elektrolyten) sowie von normalen aliphatischen Alkoholen auf die Unterdrückung der Blasenkoaleszenz in wäßrigen Lösungen zu gewinnen, wurden Absorptionsmessungen im System reines N_2/Lösung nach der „Manometrischen Methode" durchgeführt. In den Abb. 1 und 2 sind die Ergebnisse dieser Messungen in der Form $m = f(c)$ dargestellt, wobei mit m der sogenannte Beschleunigungsfaktor (enhancement factor) bezeichnet wird, der angibt, mit welchem Wert der Transportkoeffizient k_La für reines Wasser multipliziert werden muß, um den Transportkoeffizienten in der betreffenden Lösung zu erhalten:

$$m = (k_La)_{\text{Lösung}}/(k_La)_{\text{Wasser}}$$

Aus den Abb. 1 und 2 geht hervor, daß m bei intensiver Begasung (feinste Primärgasblasen) Werte bis $m = 6-7$ annehmen kann, wozu je nach Art des Salzes bzw. des Alkohols verschiedene Konzentrationen notwendig sind. Es konnte gezeigt werden [26], daß sich die Meßergebnisse an beiden Stoffklassen mit den entsprechenden $d\sigma/dc$-Werten eindeutig korrelieren lassen; die Hauptaussage der genannten Arbeit bestand jedoch im Nachweis, daß diese Korrelation auch mit Stoffparametern herbeigeführt

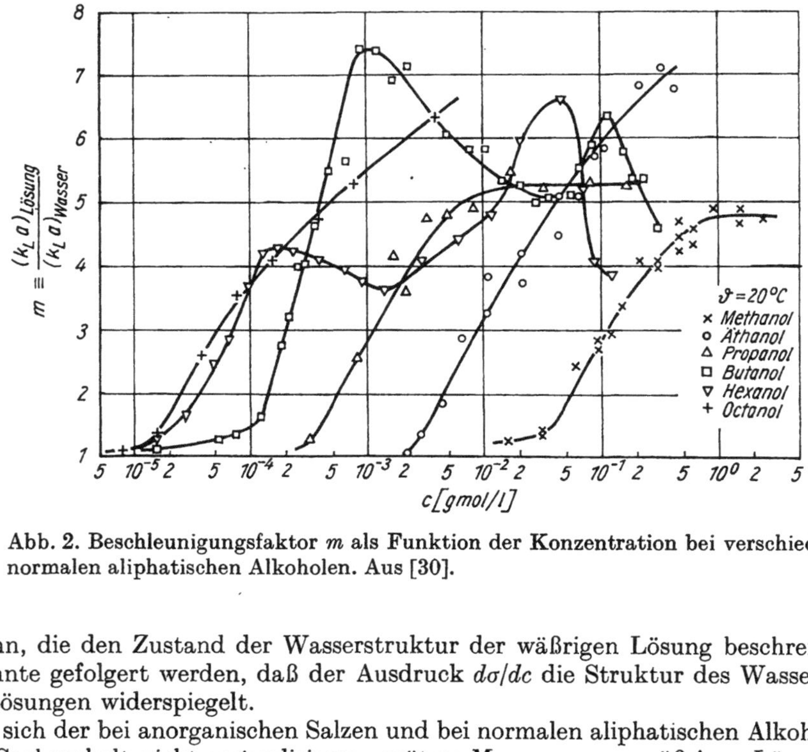

Abb. 2. Beschleunigungsfaktor m als Funktion der Konzentration bei verschiedenen normalen aliphatischen Alkoholen. Aus [30].

werden kann, die den Zustand der Wasserstruktur der wäßrigen Lösung beschreiben. Daraus konnte gefolgert werden, daß der Ausdruck $d\sigma/dc$ die Struktur des Wassers in wäßrigen Lösungen widerspiegelt.

Leider läßt sich der bei anorganischen Salzen und bei normalen aliphatischen Alkoholen gefundene Sachverhalt nicht generalisieren: spätere Messungen an wäßrigen Lösungen des Benzols und seiner Derivate zeigten komplizierte $m(c)$-Abhängigkeiten, die sich mit den zugehörigen Ausdrücken $d\sigma/dc$ allein nicht korrelieren lassen.

Es gibt jedoch auch Stoffe, die bereits in minimaler Konzentration im Wasser eine starke Beschleunigung der Koaleszenz hervorrufen. Abb. 3 zeigt Meßergebnisse mit sogenannten nicht-ionischen Tensiden, die üblicherweise als Entschäumer eingesetzt werden. Man erkennt aus dieser Darstellung, daß bei bestimmten Entschäumertypen bereits 3 ppm genügen, um den O_2-Eintrag gegenüber dem in Wasser zu halbieren. Dieser Sachverhalt macht deutlich, daß man auf den Einsatz von nicht-ionischen Tensiden zur Schaumbekämpfung verzichten sollte.

Die in den Abb. 1–3 vorgestellten Zusammenhänge belegen eindeutig, daß die (oft praktizierte) Zerlegung des Transportkoeffizienten k_La in k_L und a und eine getrennte

Verfolgung der Abhängigkeiten $k_L = f$ (Blasendurchmesser, Stoffparameter) sowie $a = f$ (Stoff- und Prozeßparameter) nur von akademischem Interesse (Transparentmachung des Vorganges) sein kann: Bestimmt man z. B. a nach der Co^{2+}-katalysierten Sulfit-Methode als Funktion des Apparatetyps und der volumenbezogenen Dispergierleistung [12], so lassen sich aus diesem Zusammenhang k_La-Werte für ein gegebenes Stoffsystem nicht ermitteln, weil man über die zugehörigen k_L-Werte nicht verfügt. Eine theoretische Berechnung von k_La [8] wird so lange eine reine Spekulation bleiben, bis die Koaleszenzphänomene restlos geklärt und vorausberechenbar geworden sind.

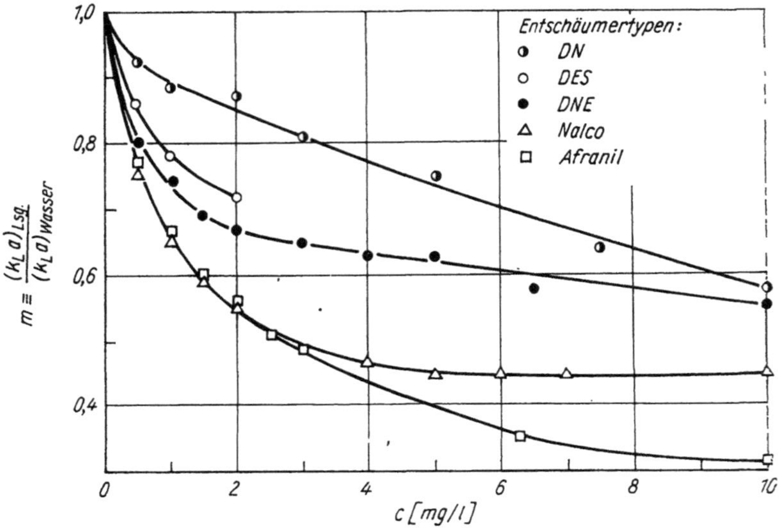

Abb. 3. Beschleunigungsfaktor m als Funktion der Konzentration bei verschiedenen Entschäumertypen. Aus [28].

Sorptions-Charakteristiken für Rührbehälter

Ähnlichkeitstheoretische Formulierung

Jahrzehntelang war es üblich, die im gegebenen Stoffsystem gefundenen k_La-Werte dimensionsbehaftet im Rahmen

$$k_La = f(P/V, v \equiv q/F) \tag{4}$$

zu korrelieren. K. VAN'T RIET [19] kommt beim Auswerten von 76 Originalarbeiten zu folgenden Zahlenwertgleichungen:

koaleszierendes System $\quad k_La = 2{,}6 \cdot 10^{-2}(P/V)^{0,4}\, v^{0,5}$

nicht-koaleszierendes System $\quad k_La = 2{,}0 \cdot 10^{-3}(P/V)^{0,7}\, v^{0,2}$

mit Dimensionen k_La [s]; P/V [W/m³]; v [m/s]; v — Leerrohrgeschwindigkeit. Demgegenüber stellt der Zusammenhang [24]:

$$k_La = f(P/q, q/V, g) \tag{5}$$

aus zweierlei Gründen eine Verbesserung dar: Zum einen ist hier der Einfluß der Erdbeschleunigung g auf ein System mit extremen Dichteunterschieden nicht vernachlässigt, zum anderen verteilt ein gut funktionierendes Rührorgan den Gasdurchsatz q im

Volumen, wodurch sich das Beziehen des Gasdurchsatzes auf den Behälterquerschnitt als unzutreffend erweist. Es läßt sich zeigen, daß die obige 4parametrige Abhängigkeit auf eine 3parametrige reduziert werden kann

$$k_L a V/q \equiv G/(q \Delta c_m) = f(P/q, g),$$

womit dann allgemein formuliert der Zusammenhang folgt:

$$k_L a V/q = f(\varrho, \nu, \sigma, S_i, D; P/q, g), \tag{6}$$

welcher in ähnlichkeitstheoretischer Schreibweise zu

$$\frac{G}{q \Delta c_m} = \varphi \left\{ \frac{P/q}{\varrho (\nu g)^{2/3}}, \frac{\nu}{D}, \frac{\sigma}{\varrho (\nu^4 g)^{1/3}}, S_i^* \right\} \tag{7}$$

führt und wie folgt abgekürzt werden kann:

$$Y_R = \varphi \{X, Sc, \sigma^* \equiv (\mathrm{Re}^4\, \mathrm{Fr})^{1/3}\, \mathrm{We}^{-1}, S_i^*\} \tag{8}$$

In diesem Rahmen ausgewertete $k_L a$-Messungen führen zur Aufstellung von Sorptions-Charakteristiken für Rührbehälter.

Koaleszierendes Stoffsystem

Für dieses Stoffsystem liegen viele Meßergebnisse vor. Im Gegensatz zu eigenen Untersuchungen [24] sowie jenen von Uhl u. a. [18] fanden Yagi u. Yoshida, Höcker sowie Kipke niedrigere und somit verläßlichere $k_L a$-Werte; diese wurden kürzlich von Henzler [5] vergleichend dargestellt, Abb. 4. Auch Ergebnisse von Grasshoff [4] sowie Smith u. a. [16] passen gut in dieses Bild.

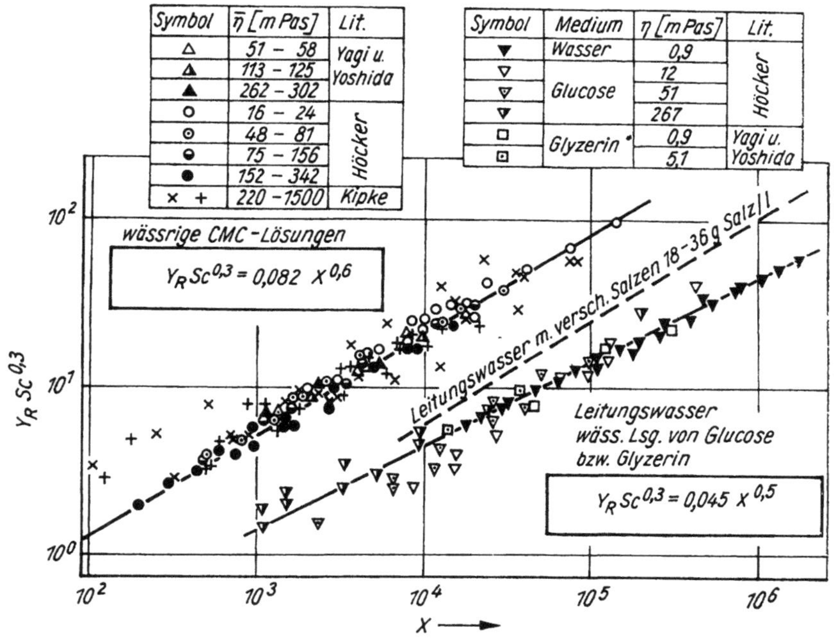

Abb. 4. Sorptions-Charakteristiken eines Rührbehälters mit Scheibenrührer für verschiedene Stoffsysteme. Korrelation von Henzler [5]; die gestrichelte Gerade gibt die Meßergebnisse aus [14] und [16] wieder.

Nicht-koaleszierendes Stoffsystem

Für dieses Stoffsystem liegen seit COOPER u. a. [1] unzählige Meßergebnisse vor (s. auch [19]), die alle im Natriumsulfit-System ermittelt wurden. Leider sind die Literaturangaben für eine Übertragung in den vorgeschlagenen Auswertungsrahmen zu dürftig. Nur aus diesem Grund können hier ausschließlich eigene Versuchsergebnisse [25] vorgestellt werden, die mit einer 1 nNa_2SO_3-Lösung und selbstansaugenden Hohlrührern bei Messungen im Maßstab 1 : 5 gewonnen wurden, vgl. Abb. 5.

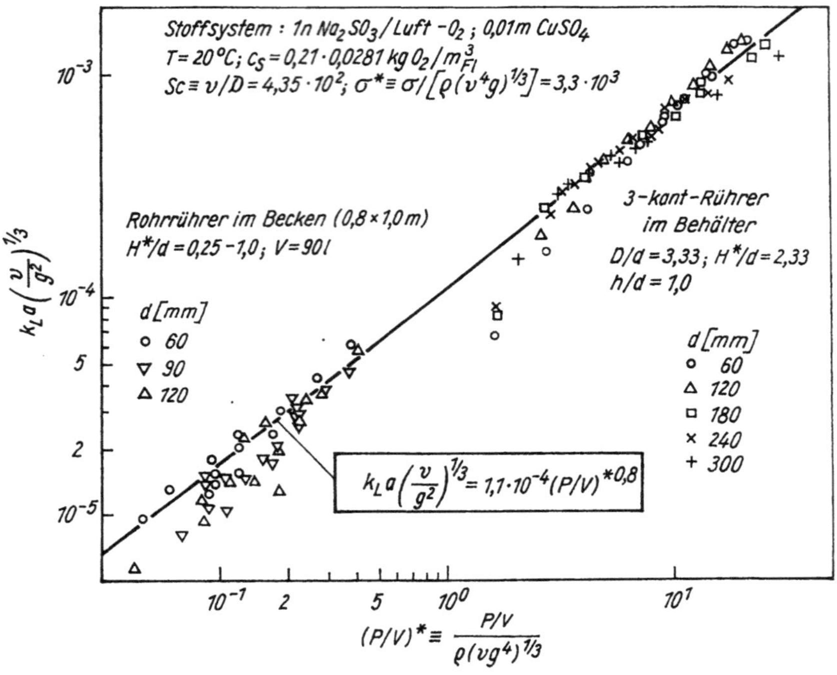

Abb. 5. Sorptions-Charakteristik eines Behälters mit Hohlrührer im vollständig koaleszenzgehemmten Stoffsystem. Aus [25].
D, d — Behälter- und Rührerdurchmesser; H, h — Flüssigkeitsüberdeckung und der Bodenabstand des Rührers. $v \triangleq$ kinematische Viskosität des Wassers ν (gilt nicht für $Sc \equiv$)

Der in Abb. 5 eingerahmte analytische Ausdruck läßt sich umformen zu

$$Y_R = 1{,}1 \cdot 10^{-4} X^{0{,}8} (q/V)^{*-0{,}2},$$

wobei

$$(q/V)^* \equiv (q/V)\,(\nu/g^2)^{1/3}$$

bedeutet. Auch die anfangs erwähnte dimensionsbehaftete Korrelation für $k_L a$ in nicht-koaleszierenden Systemen (vgl. [19]):

$$k_L a \doteq 2{,}0 \cdot 10^{-3} (P/V)^{0{,}8}\, v^{0{,}2}$$

läßt sich mit dem vorgestellten, für koaleszierende Stoffsysteme zutreffenden ähnlichkeitstheoretischen Rahmen nur nach Hinzunahme einer weiteren Kennzahl formulieren.

Sie lautet umgeformt

$$Y_R = 1{,}46 \cdot 10^{-4}\, X^{0{,}8}\, H^{*0{,}2}$$

wobei

$H^* \equiv H(g/\nu^2)^{1/3}$ ist und $H/D = 1$ angenommen wurde.

Sorptions-Charakteristiken für Blasensäulen

Fritte oder Lochboden als Gaszerteiler

Bei Blasensäulen mit Gaszerteilern wie Fritten- oder Lochböden ist der Gasdurchsatz der einzige Prozeßparameter und für ein gegebenes Stoffsystem kann der Zusammenhang

$$k_L a = f(v, g)\,; \quad v - \text{Leerrohrgeschwindigkeit} \tag{9}$$

erwartet werden. Tatsächlich wird oft gefunden (vgl. z. B. [15]), daß $k_L a \sim v$ ist. Dieser Tatbestand wird durch folgende Umformung [28] klarer:

$$\frac{k_L a}{v} \equiv \frac{G}{V \Delta c_m} \cdot \frac{F}{q} = \frac{G}{H q \Delta c_m} = \text{const}\,; \tag{10}$$

sie verdeutlicht, daß hier der O_2-Eintrag G der Flüssigkeitshöhe H, dem Gasdurchsatz q und der Konzentrationsdifferenz Δc direkt proportional ist. Die aus diesem Ausdruck gewonnene Sorptions-Kennzahl

$$Y_B \equiv \frac{k_L a}{v} \left(\frac{\nu^2}{g}\right)^{1/3} \equiv \frac{G}{H q \Delta c_m} \left(\frac{\nu^2}{g}\right)^{1/3}$$

wird so lange einen konstanten Zahlenwert behalten, solange $k_L a \sim v$ bleibt.
Es ist zu erwarten, daß Y_B bei koaleszierenden Stoffsystemen mit steigendem Gasdurchsatz (bzw. Leerrohrgeschwindigkeit v) abnehmen wird, weil hohe Begasungsdichte die Blasenkoaleszenz begünstigt. Bei nicht-koaleszierenden Stoffsystemen wird dagegen Y_B mit steigendem v^* eher größere Werte annehmen, weil höhere Gasdurchsätze im Lochboden höhere Scherkräfte und damit feinere Primärgasblasen bewirken, die im nicht-koaleszierenden Stoffsystem konserviert bleiben. Auch bei solchen Sachverhalten ($k_L a \not\sim v$) empfiehlt es sich, Y_B als Kennzahl zu behalten; sie ist von hier $v^* = v(\nu g)^{-1/3}$ abhängig: $Y_B(v^*)$.
Abb. 6 zeigt zunächst die gemessenen Zusammenhänge $k_L a(v)$ aus [15] für eine Blasensäule mit Lochboden (A) bzw. Sinterplatte (B) in Abhängigkeit vom Koaleszenzverhalten des Stoffsystems und dann die Darstellung dieser Ergebnisse in der besprochenen Weise als $Y_B(v^*)$. In dieser Darstellung erkennt man deutlich den großen Einfluß der Koaleszenzhemmung durch Alkoholzusätze. Während bei der Sinterplatte Y_B im großen v^*-Bereich konstant bleibt, steigt Y_B beim Lochboden leicht mit v^* an (bei Wasser A 1 : $Y_B \sim v^{*0{,}28}$ und bei Alkohollösungen A 2 : $Y_B \sim v^{*0{,}40}$). Diese Abhängigkeit ist in Abb. 7 noch stärker ausgeprägt ($Y_B \sim v^{*0{,}5}$), weil hier eine 1 n Na_2SO_3-Lösung (vollständig unterdrückte Blasenkoaleszenz) vorlag; die Abb. 7 zeigt, daß auch bei einem solchen Stoffsystem mit steigender Flüssigkeitshöhe (längere Verweilzeit der Gasblasen) die Koaleszenz zunimmt. Die in Abb. 7 vorgestellten Meßergebnisse sind an drei Blasensäulen mit Lochboden (Bohrungs-⌀ 0,5 mm) gewonnen worden [22]. Abb. 8 zeigt die Ergebnisse von Messungen in hohen Blasensäulen, bei denen die Flüssigkeit (Leitungswasser) im Gleichstrom bzw. Gegenstrom zu Gasblasen strömte [3]. In unserer Schreib-

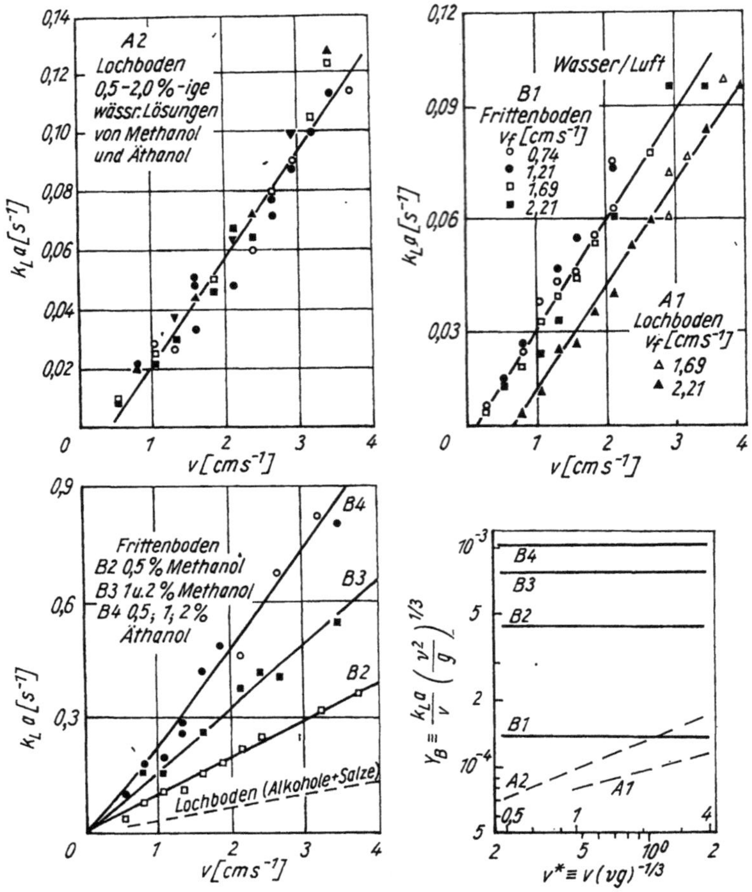

Abb. 6. Dimensionsbehaftete Darstellung $k_L a(v)$ für eine Blasensäule (0,14 ⌀ × 4,0 m), mit Sinterplatte bzw. Lochboden als Gaszerteiler in Abhängigkeit vom Koaleszenzgrad des Systems aus [15] und die übersichtlichere dimensionslose Darstellung $Y_B(v^*)$.

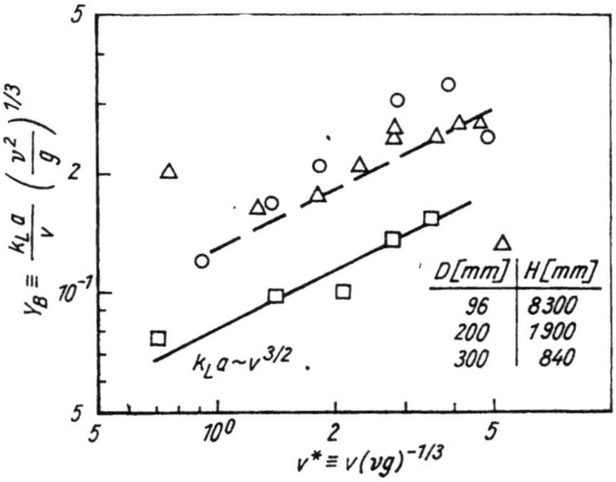

Abb. 7. Abhängigkeit $Y_B(v^*)$ für drei Blasensäulen mit Lochboden (Bohrungen von 0,5 mm ⌀) bei vollständig koaleszenzgehemmtem Stoffsystem. Nach [22].

Gaszerteiler	d_c [cm]	L [cm]	Fluß	Symbol
Begasungs-kreuz, Bohr-ungen 1 mm	20	250	↑↑	×
	20	723	↑↑	○
	20	723	↑↓	□
Sinterplatte 150 µm	10	250	↑↑	●
	15	440	↑↑	△
	15	440	↑↓	◇

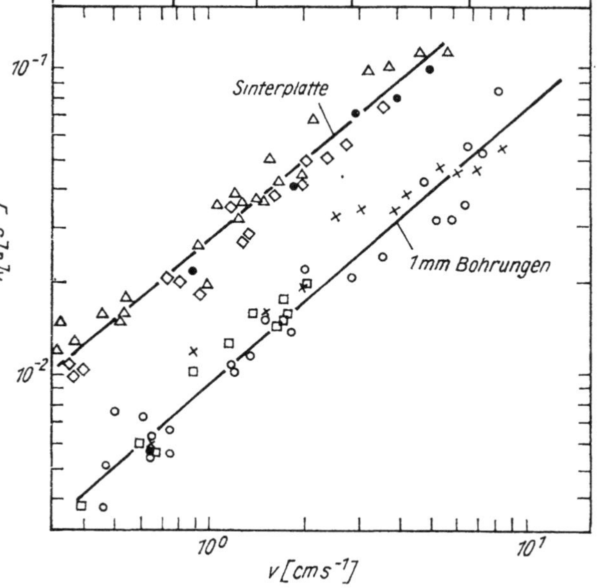

Sinterplatte: $Y_B = 1{,}17 \cdot 10^{-4}\ v^{*-0{,}1}$

1mm-Bohrungen: $Y_B = 3{,}9 \cdot 10^{-5}\ v^{*-0{,}1}$

Abb. 8. $k_La(v)$-Abhängigkeit bei Blasensäulen mit verschiedenen Gaszerteilern und Führung des Flüssigkeitsdurchsatzes (reines Wasser) im Gleich- bzw. Gegenstrom zum Gasdurchsatz. Aus [3].

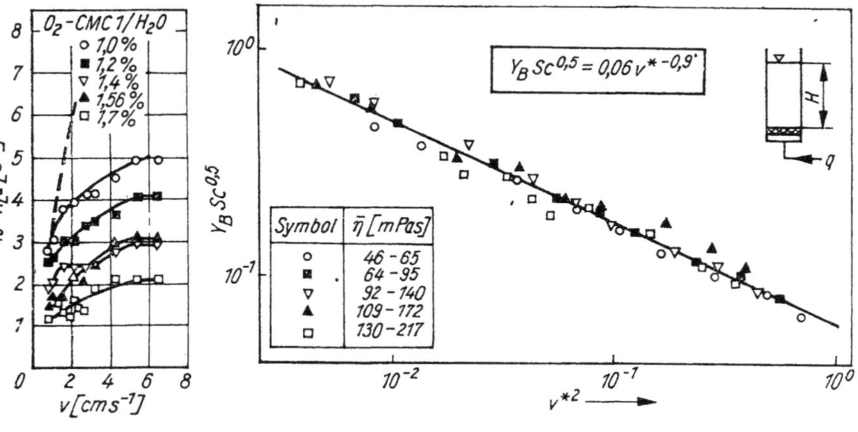

Abb. 9. Dimensionsbehaftete und dimensionslose Darstellung des Stofftransportes in einer Blasensäule (0,14 ⌀ × 3,91 m) mit Sinterplatte bei wäßrigen CMC-Lösungen. Aus [5].

weise lauten die Sorptions-Charakteristiken

für die Sinterplatte: $Y_B = 1{,}17 \cdot 10^{-4} v^{*-0,1}$

und für den Lochboden: $Y_B = 3{,}9 \cdot 10^{-5} v^{*-0,1}$.

Bei strukturviskosen (nicht-NEWTONschen) Fluiden kann die Koaleszenz mit steigendem Gasdurchsatz dermaßen gefördert werden, daß $k_L a$ von v fast unabhängig wird, somit wird $Y_B \sim 1/v$, vgl. Abb. 9.

Injektoren als Gaszerteiler

Bei Stoffsystemen mit unterdrückter Koaleszenz lohnt es sich, besonders feine Gasblasen zu erzeugen. Dazu eignen sich vor allem Injektoren, bei denen die kinetische Energie des Flüssigkeitstreibstrahles zum Zerteilen des Gases verwendet wird. Wegen

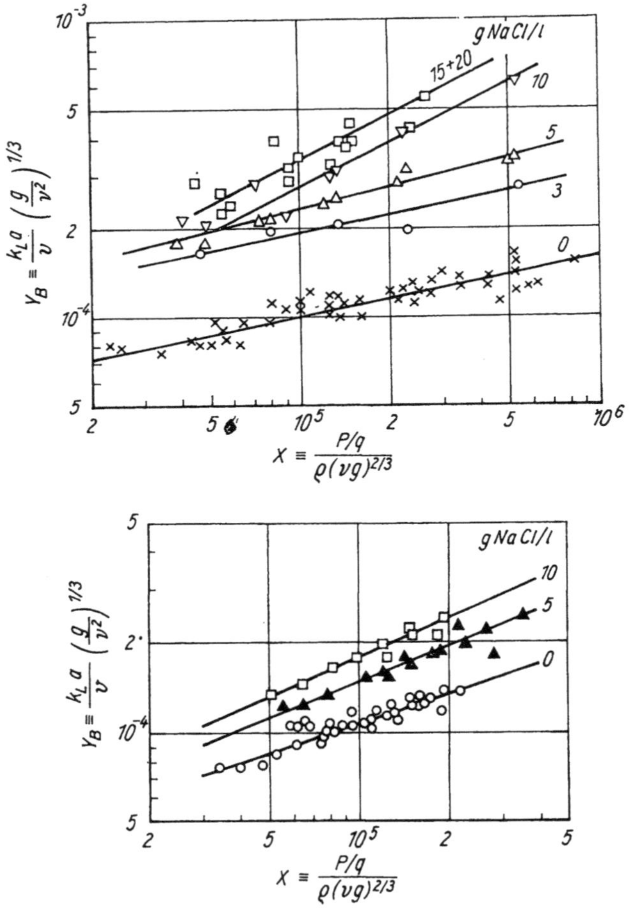

Abb. 10. Sorptions-Charakteristiken $Y_B(X)$ zweier Injektortypen in Abhängigkeit des Koalreszenzgrades des Systems, ausgedrückt durch die NaCl-Konzentration.
Oben: BAYER-Injektor 8/14 im Behälter 1,6 ⌀ × 8 m;
unten: BAYER-Schlitzstrahler im Behälter 2,8 ⌀ × 8 m.
Aus [27]. $\left(X \equiv \frac{P/q}{\varrho(\nu g)^{2/3}}\right)$

des zweiten Prozeßparameters (Flüssigkeitsdurchsatz) muß der ähnlichkeitstheoretische Rahmen um eine Kennzahl erweitert werden. Es wurde vorgeschlagen [28], als solche nicht das Durchsatzverhältnis q_n/q_f zu wählen, sondern die gleiche Kennzahl X wie bei den Rührern zu verwenden, die hier mit der Treibstrahlleistung des Flüssigkeitsstrahles $P_f = \Delta p q_f$ zu bilden ist. Somit ist im allgemeinen Fall die Sorptions-Charakteristik für Zweistoffdüsen (Injektoren, Ejektoren, Venturi-Düsen usw.) im Rahmen

$$Y_B = f(X, Sc, \sigma^*, S_i^*)$$

wiederzugeben.

Bislang sind die Sorptions-Charakteristiken nur für zwei Injektortypen publiziert worden [27], nämlich für den BAYER-Injektor 8/14 sowie für den BAYER-Schlitzstrahler. Die Messungen sind mit Vorrichtungen technischer Größe in entsprechend großen Blasensäulen ($D = 1{,}6$ und $2{,}8$ m, $H = 7$ m) durchgeführt worden, wobei das Koaleszenzverhalten des Systems durch NaCl-Zugabe variiert wurde. Die dabei gewonnenen Ergebnisse sind in den Sorptions-Charakteristiken in Abb. 10 wiedergegeben. Man erkennt, daß die Sorptions-Charakteristiken beim kleineren Injektor 8/14 im wesentlich stärkeren Maße von der Salzkonzentration abhängen als beim Schlitzstrahler. Das beruht darauf, daß der Injektor 8/14 feinere Primär-Gasblasen als der Schlitzstrahler erzeugt.

Dimensionierung und Optimieren von Rührbehältern und von Blasensäulen für den gewünschten O_2-Eintrag

Die in diesem Aufsatz vorgestellten Sorptions-Charakteristiken für Rührbehälter und Blasensäulen erlauben eine Dimensionierung und Optimierung von entsprechenden Apparaten, soweit das zu behandelnde Stoffsystem in bezug auf sein Koaleszenzverhalten und seine effektive Viskosität dem Versuchs-Stoffsystem entspricht. Oft wird man finden, daß dies nicht der Fall ist. Dann sind Labormessungen mit zutreffendem (Modell-)Stoffsystem notwendig. Dazu muß betont werden, daß wegen des Antagonismus zwischen der Blasenzerteilung und Blasenkoaleszenz Laborversuche nur bedingt eine Auskunft über die Gegebenheiten in der technischen Anlage geben können. Je koaleszierender das Stoffsystem, desto wichtiger ist es, Versuche an einer möglichst großen halbtechnischen Anlage durchzuführen, vgl. dazu auch [27].

Zur Auslegung und Optimierung von *Rührbehältern* als Fermentor ist außer den Sorptions-Charakteristiken die Kenntnis der sogenannten Überflutungs-Charakteristiken und der Leistungs-Charakteristiken im Stoffsystem Gas/Flüssigkeit notwendig. Ausführliche Angaben hierzu sind in [21, 23] und [6] enthalten.

Je höher der notwendige Stofftransport, desto höher kann auch der erforderliche Wärmeaustausch sein [10]. Umfassende Auslegungswerte über den Wärmeübergang in Rührbehältern bei unbegasten Flüssigkeiten sind in [13] zusammengestellt. Durch Begasung wird der Wärmeaustausch zwischen Flüssigkeit und Behälterwand bei gutem Rühren nur geringfügig verschlechtert, bei schwachem Rühren jedoch wesentlich verbessert [17].

Zur Auslegung und Optimierung einer *Blasensäule* ist außer der Sorptions-Charakteristik die Kenntnis des Gasvolumens in der Flüssigkeit ("gas hold up") notwendig, damit der Apparat für das gewünschte Flüssigkeitsvolumen richtig bemessen werden kann. Die Angaben über diese leicht zu messende Größe sind als sogenannte „Sprudelschichterhöhung“ h oder „scheinbare Dichte“ ε in fast allen Arbeiten enthalten, die sich mit Blasensäulen beschäftigen. Es wird nur allzu oft vergessen, daß diese Größe wegen der Koaleszenzphänomene nicht nur extrem stark stoffsystemabhängig ist, sondern auch

von den geometrischen Parametern des Apparates (Durchmesser, Flüssigkeitshöhe) abhängt und somit maßstabsabhängig ist.
Der erforderliche Wärmeübergangskoeffizient in Blasensäulen ist den Arbeiten von KAST [7] und BURKEL [2] zu entnehmen. Der Druckverlust eines Lochbodens [20] oder des sonstigen Gaszerteilers ist notwendig, um die adiabatische Verdichterleistung für das Gas berechnen zu können, wobei die Wirkungsgerade des Gasverdichters ($\eta = 0{,}60$) und bei Injektoren der Flüssigkeitspumpe ($\eta = 0{,}75$) berücksichtigt werden müssen, damit verbindliche Aussagen über die Effizienz [kg O_2/kWh] der Blasensäule gewonnen werden können, vgl. z. B. [27].
Die Voraussagen für den volumenbezogenen O_2-Eintrag G/V sowie die dabei erreichbare Effizienz E sind für eine Reihe von verschiedenen Belüftungsapparaten in [29], für Blasensäulen mit Injektor als Gaszerteiler in [27] sowie für Rührer in [29] und [5] enthalten.

Literatur

[1] COOPER, C. M., FERNSTROM, G. A., MILLER, S. A.: Performance of Agitated Gas-Liquid Contactors; Ind. Eng. Chem. **36** (1944) 6, 504.

[2] BURKEL, W.: Der Wärmeübergang an Heiz- und Kühlflächen in begasten Flüssigkeiten; Chem.-Ing.-Techn. **44** (1972) 5, 265.

[3] DECKWER, W. D.: Physical Transport Phenomena in Biological Reactors; Proc. 6th Int. Fermentation Symp., London/Ont., Canada, 21.–25. July 1980.

[4] GRASSHOFF, A.: Über den Eintrag von Sauerstoff in wäßrige Fermentationsmedien mit Hilfe selbstansaugender Begasungsrührer; Dissertation TU Berlin 1978.

[5] HENZLER, H. J.: Begasen höherviskoser Flüssigkeiten; Chem.-Ing.-Techn. **52** (1980) 8, 643.

[6] JUDAT, H.: Eignung schnellaufender Rührertypen zum Dispergieren von Gasen in niedrigviskosen Flüssigkeiten; Chem.-Ing.-Techn. **48** (1976) 3, 228.

[7] KAST, W.: Untersuchungen zum Wärmeübergang in Blasensäulen; Chem.-Ing.-Techn. **35** (1963) 11, 785.

[8] KUDREWIZKI, F.: Modellierung des Stoffüberganges Gas–Flüssigkeit in Rührkesseln; Chem. Techn. **32** (1980) 6, 302.

[9] LOPES DE FIGUEIREDO, M. M. und CALDERBANK, P. H.: The Scale-up of Aerated Mixing Vessels for specified Oxygen Dissolution Rates; Chem. Eng. Sci. **34** (1979), 1333.

[10] LUONG, J. H. T. and VOLESKI, B.: Determination of the Heat of some Aerobic Fermentations; Canad. J. of Chem. Eng. **58** (1980), 497.

[11] MARUCCI, G.: A Theory of Coalescence; Chem. Eng. Sci. **24** (1969), 975.

[12] NAGEL, O., KÜRTEN, H. und SINN, R.: Stoffaustauschfläche und Energiedissipationsdichte als Auswahlkriterien für Gas/Flüssigkeits-Reaktoren; Chem.-Ing.-Techn. **44** (1972) 14, 899.

[13] POGGEMANN, R. u. a.: Wärmeübergang in Rührkesseln mit einphasigen Flüssigkeiten; Chem. Ing.-Techn. **10** (1979) 10, 948

[14] ROBINSON, C. W. und WILKE, C. R.: Oxygen Absorption in Stirred Tanks: A Correlation for Ionic Strength Effects; Biotechn. and Bioeng. **15** (1973), 755.

[15] SCHÜGERL, K., LÜCKE, J. und OELS, U.: Bubble Column Bioreactors; Advances in biochem. Engng. **7** (1977), 1–84.

[16] SMITH, J. M., VAN'T RIET, K. und MIDDLETON, J. C.: Scale-up of Agitated Gas-Liquid Reactors for Mass Transfer; Paper F4 on the Second European Conference on Mixing, GB-Cambridge, March 1977.

[17] STEIFF, A. und WEINSPACH, P. M.: Wärmeübergang in gerührten und ungerührten Gas/Flüssigkeits-Reaktoren; Chem.-Ing.-Techn. **49** (1977) 2, 157, Synopse 446/77.

[18] UHL, V. W., WINTER, R. L. und HEIMARK, E. L.: Mass Transfer in Large Secondary Treatment Aerators Water-1976, AIChE Symposium Series, No. 167, Vol. 73, 33.

[19] VAN'T RIET, K.: Review of Measuring Methods and Results in Nonviscous Gas-Liquid Mass Transfer in Stirred Tanks; Ind. Eng. Chem. Process Des. Dev. **18** (1979) 3, 357.

[20] ZELFEL, E.: Der Druckverlust beim Durchströmen der Flüssigkeitsschicht auf einem Siebboden; Chem.-Ing.-Techn. **39** (1967) 5/6, 284

[21] ZLOKARNIK, M. und JUDAT, H.: Rohr- und Scheibenrührer — zwei leistungsfähige Rührer zur Flüssigkeitsbegasung; Chem.-Ing.-Techn. **39** (1967) 20, 1163.
[22] ZLOKARNIK, M.: Eignung von Einlochböden als Gasverteiler in Blasensäulen; Chem.-Ing.-Techn. **43** (1971) 6, 329.
[23] ZLOKARNIK, M.: Rührleistung in begasten Flüssigkeiten; Chem.-Ing.-Techn. **45** (1973) 10a, 689.
[24] ZLOKARNIK, M.: Einfluß einiger stofflicher und verfahrenstechnischer Parameter auf den Sauerstoffeintrag bei der Abwasserbelüftung; Chem.-Ing.-Techn. **47** (1975) 7, 281
[25] ZLOKARNIK, M.: Sorptions Characteristics for Gas-Liquid Contacting in Mixing Vessels; Advances in Biochem. Engng. 8 (1978), 133.
[26] ZLOKARNIK, M.: Influence of various Material and Process Related Parameters on Bubble Coalescence in Gas/Liquid Contacting; Discussion paper 1/13 on the First European Congress on Biotechnology, CH-Interlaken, September 1978.
[27] ZLOKARNIK, M.: Die Leistungsfähigkeit optimierter Injektoren für den Sauerstoffeintrag in biologische Abwasserreinigungsanlagen; Verfahrenstechnik **13** (1979) 7/8, 601.
[28] ZLOKARNIK, M.: Sorption Characteristics of Slot Injectors and their Dependency on the Coalescence Behaviour of the System; Chem. Eng. Sci. **34** (1979) 10, 1265.
[29] ZLOKARNIK, M.: Eignung und Leistungsfähigkeit von Volumenbelüftern für biologische Abwasserreinigungsanlagen; Korrespondenz Abwasser **27** (1980) 3, 194.
[30] ZLOKARNIK, M.: Koaleszenzphänomene im System gasförmig/flüssig und deren Einfluß auf den O_2-Eintrag bei der biologischen Abwasserreinigung; Korrespondenz Abwasser **27** (1980) 11 728.

BOOK REVIEW

U. E. VIESTURS, M. Z. KRISTAPSONS, E. S. BYLINKINA

Cultivation of Microo-ganismus

Moscow, Food Industry, 1980, (russ.) 232 pages

The rapid development of microbiological production recently has created the necessity to summarize bioengineering problems scattered in periodical editions. The book under review generalizes and analysis the information of periodical editions published up to 1978. The book also makes use of the experience and the results of personal experiments of the authors obtained at the Institute of Microbiology named after A. KIRHENŠTEINS of the Latvian SSR Academy of Sciences (U. E. VIESTURS, M. Ž. KRISTAPSONS) and at the All-Union Scientific Research Institute of Antibiotics of the USSR Ministry of Medical Industry (E. S. BYLINKINA).

Chapter I deals with pre-fermentation processes: preparation and sterilization of feed media, purification and sterilization of air, ways of obtaining the inoculum.

Chapter 2 is devoted to questions of aeration, mixing and scaling. Questions of oxygen and carbon dioxide mass transfer have been treated in detail, the hydrodynamic conditions in the fermentor have been evaluated. Many assuptions have been based on the results of the authors' personal research.

Chapter 3 deals with the role of foaming in the process of submerge cultivation of microorganisms. As a result of the analysis of chemical foam suppressors and mechanical devices for foam suppression, the authors provide a criterion for the evaluation of the efficiency of foam suppression regarding the specific liquid escape from the fermentor, as well as give recommendations for chemico-mechanical method of foam suppression. The control of this process is carried out via special electronic device, designed by the authors.

Chapter 4 describes the basic equipment for microorganism cultivation — fermentors, beginning with laboratory ones up to industrial-scale fermentors, i.e., with the volume from 2 l up to 100 m^3 and more. Fermentors with various design solutions, with the energy introduced by aerating gas and those with mechanical mixing are compared. Fermentors of individual design have been tested under production conditions, according to the results of lysine biosynthesis by bacteria.

Chapter 5 includes data on the collection of forst-hand information, automatic control and guiding the processes of microbial biosynthesis. Sensors of primary information (pH, Eh, pO_2, pCO_2, biomass, indication of foam, ferment electrodes etc.) are briefly dealt with. Principles of automatic regulation of technological parameters, starting with local regulators and ending with the use of computers in the control systems according to mathematical models have been given.

In the 6th, the concluding chapter, variants for processing of the products of microbial synthesis have been listed: of cell mass, concentrates, intercellular and extracellular metabolites.

The present work reviews all the basic problems connected with the cultivation of microorganisms. The book carries a subheading "Foundations of Bioengineering", — evidently this is why the authors have excluded from the range of reviewed problems the biological aspects of microorganism cultivation: obtaining of active producents, the bases for regulating metabolism of producents and others. It is difficult within the scope of a single book to embrace all the problems of microorganism cultivation fully enough. Most informative are Chapters 2, 3, 4 which deal with questions connected with the scientific activity of the authors. Chapters 5 and 6 are most conspective, especially brief in the description of automation and optimization principles of cultivation processes. The contents of the book shows that the authors have used a host of literature sources, however, the number of references at the end of each chapter is not sufficient. The book is concluded by a subject index which makes it easy for reader to find the required material and use the presented reference data.

On the whole, all the chapters reflect the current state of the problem and the book will serve as a good manual for workers of various qualification taking interest in the engineering aspects of microorganism cultivation.

J. E. ŠVINKA

Acta Biotechnologica 1 (1981) 4, 327–337

The Role of Mechanical Shear and Cavitation on the Behaviour of Molds

J. PLACEK, E. UJCOVÁ, M. MUSÍLKOVÁ, L. SEICHERT and Z. FENCL

Institute of Microbiology, Czechoslovak Academy of Sciences
Videnska 270, 142 20 Prague 4, CSSR

Paper given at the 2nd Symposium of Socialist Countries on Biotechnology, Leipzig 2.–5. 12. 1980

Summary

The morphology of filamentous microorganisms does essentially affect the production of metabolites. Agitating conditions may affect the morphology and for this reason the production of metabolites too.
The following parameters it was found to have an influence:

Reynolds mixing number
impeller blade tips velocity
mean shear stress close to the impeller
impeller power consumption per unit volume
cavitation pressure drop

It were presumed three mechanisms for the mechanical effect on the microorganisms:

1. the direct impact of the impeller blades on the microorganisms-collision
2. the shear stress in the liquid phase
3. a sharp pressure decrease behind the impeller blades-cavitation

Mathematical relationships are developed for the different mechanisms.
Using *Aspergillus niger* it is shown what morphological and physiological states of this microorganism are caused by mechanical straining and the conditions for the maximal production of citric acid are studied.
Requirements for scale-up are discussed.

Zusammenfassung

Die Morphologie von filamentösen Mikroorganismen hat einen wesentlichen Einfluß auf die Produktion von Metaboliten. Rührbedingungen beeinflussen die Morphologie und damit die Metabolitproduktion.
Als wesentliche Einflußgrößen werden genannt:

Re-Zahl für Rührer
Rührerblattspitzengeschwindigkeit
mittlere Scherspannung in Rührernähe
Leistungseintrag des Rührers
Druckabfall durch Sog hinter den Rührerblättern.

Es werden drei Mechanismen des mechanischen Rühreinflusses auf die Mikroorganismen angenommen:

1. direkter Kontakt der Rührerblätter mit den Mikroorganismen — Kollision
2. Scherspannung in der flüssigen Phase
3. Druckabfall hinter den Rührerblättern — Kavitation.

Für die einzelnen Mechanismen werden mathematische Beziehungen angegeben.
Am Beispiel von *Aspergillus niger* wird gezeigt wie morphologischer und physiologischer Zustand der Kultur von den mechanischen Einwirkungen beeinflußt werden und damit die Zitronensäureproduktion.
Erfordernisse für eine Maßstabsübertragung werden diskutiert.

The design of the present-day industrial fermenters usually represents a compromise between two basic approaches — the chemical engineering one and the biological one. For a chemical engineer the fermenter design should ensure a good heat and mass transfer at a low impeller power consumption. The consequence of such an approach is the existence of two different regions in the fermenter, one in a direct vicinity of the impeller with an intensive transfer of momentum, heat and mass, and the other one created by the recirculation loop in the remaining volume of the fermenter with much lower intensities of heat and mass transfer. For a biologist, on the other hand, the main goal is both to retain the identical cultivation conditions in the scale-up process (i.e. to retain the same conditions in the laboratory and in the production fermenters), and to achieve a high degree of homogenity in the fermenter so that the culture grows under precisely defined conditions. In the single-cell protein production the biological aspects of such a process may often be neglected in comparison with the chemical engineering ones because for the usual biomass concentrations range the fermentation medium behaves like a newtonian fluid, monocellular microorganisms are little sensitive to changes in hydrodynamical conditions in the fermenter and a small change in cultivation conditions usually results only in a small biomass production change. On the other hand, in the microbial production of metabolites (enzymes, antibiotics, steroids and organic acids), even a small change in cultivation conditions may significantly affect the physiological state of microorganisms. Furthermore, due to non-newtonian behaviour of the suspensions of filamentous microorganisms the flow patterns in the cultivation apparatuses cause much worse conditions for homogenity and overall heat and mass transfer.
It has been known since the beginning of the submerged production of antibiotics that fungi and actinomycetes require a special mode of agitating in order to reach the optimal growth rate and morphology which are connected with the production phase. The results of Dion et al. [1], Gabor (Effect of shear on *Penicillin* molds, Ph. D. Thesis, Cornell University (1958)), and the communication of Camposano et al. at the VII. International Congress for Microbiology, Stockholm (1958) show that with an increasing impeller speed the pelletary form of growth changes to the diffuse one which again changes its morphology by shortening the hyphae. This brings about a radical change in the hydrodynamic conditions in the fermenter due to the changes in flow properties of the medium and the attendant changes in heat and mass transfer. A change in the typical size of aggregates formed by the filamentous microorganisms from pellets to the considerably smaller clusters of a diffusively growing mycelium causes the decreased differentiation along the filaments which has often a decisive significance for the production of metabolites. Dion et al. [1] showed that almost the same morphological changes of the mycelium can be obtained either by increasing the impeller speed or by increasing the oxygen partial pressure, and that these effects cannot be easily separated in an usual fermenter equipped with a turbine impeller providing both dispergation and pumping effects. In any case the transfer of experimental results obtained from fermentational processes

using filamentous microorganisms is still empirical because the impeller-induced effect on mycelium is unclear.

Similarly, in developing a new technology for a submerged citric acid production using *Aspergillus niger* strains we had to find cultivation conditions suitable for the production in the large-scale fermenter. Fig. 1 gives the values of some parameters at which the highest production of citric acid was achieved: the Reynolds mixing number Re_M, the impeller blade tips velocity v_{circ}, the mean shear stress close to the impeller τ, the impeller power consumption per unit volume P/V, and the cavitation pressure drop Δp. We can see that for the most efficient fermentations the values the mean shear stress τ and the specific impeller power consumption P/V decrease with increasing fermenter volume V. In other words, with an increasing fermenter volume (and with decreasing

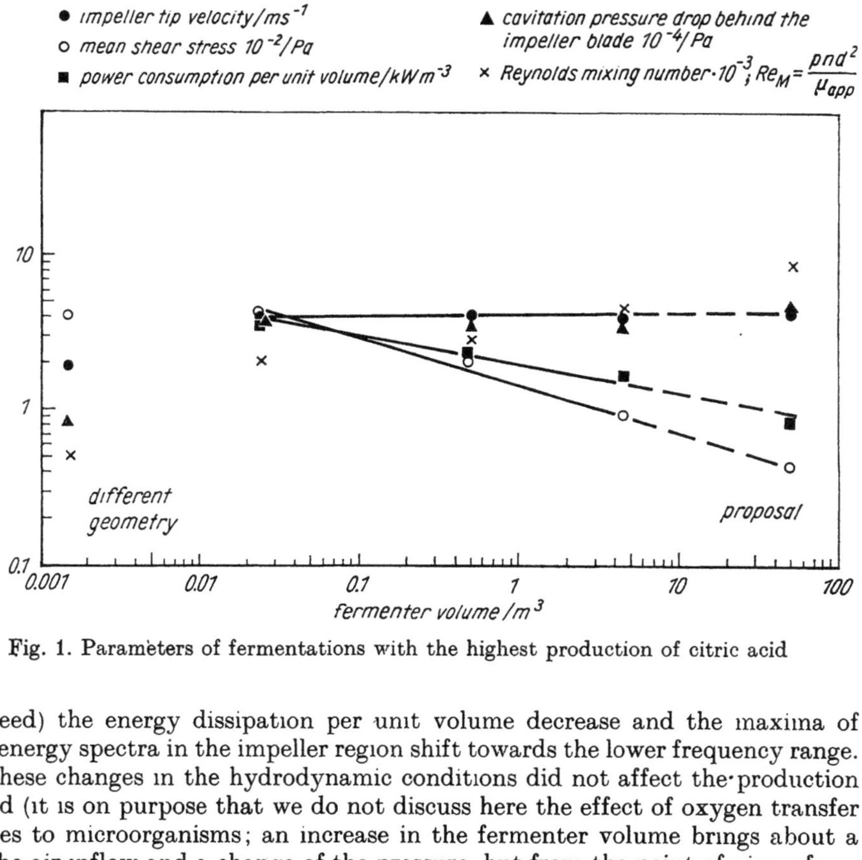

Fig. 1. Parameters of fermentations with the highest production of citric acid

impeller speed) the energy dissipation per unit volume decrease and the maxima of turbulence energy spectra in the impeller region shift towards the lower frequency range. However, these changes in the hydrodynamic conditions did not affect the production of citric acid (it is on purpose that we do not discuss here the effect of oxygen transfer from bubbles to microorganisms; an increase in the fermenter volume brings about a change in the air inflow and a change of the pressure, but from the point of view of our contribution these changes are not essential.) On the other hand, a satisfactory citric acid yield was achieved for the same values of circumferential velocity v_{circ} and its change caused a significant change in the citric acid production. Fig. 2 shows the dependence of the citric acid production on the impeller speed obtained by cultivation in the laboratory fermenter (working volume $V = 1.5$ l, turbine impeller diameter $d = 50$ mm). A marked yield maximum was observed for every tested mutant. The low yield at low impeller speeds was probably due to low mass transfer which manifested itself in a low biomass concentration. However, beyond the maximum the increasing impeller speed brought about a decrease in the citric acid yield even thought the biomass concentration still increased. The morphology of hyphae changed simultaneously with the biosynthetical activity. The changes are depicted in Fig. 3a and Fig. 3b. The filaments of *Aspergillus niger* S59 which grew under the impeller speed $n = 400$ rpm were long, thin, sparcely septated and contained only a small number of septa, while those which grew under the high impeller speed $n = 1200$ rpm were shorter, thick, more branched and contained a large number of septa.

Due to the complicated flow pattern of the suspension of microorganisms discharged from the turbine impeller the analysis of its mechanical effect on the microorganisms is very complex. According to our model which was presented at the First Symposium of the Socialist Countries on Biotechnology in Riga (1978), we presume the tree following mechanisms:

1. the direct impact of the impeller blades on the microorganisms — collision,
2. the shear stress in the liquid phase (turbulence effects fall within this category), and
3. a sharp pressure decrease behind the impeller blade — cavitation.

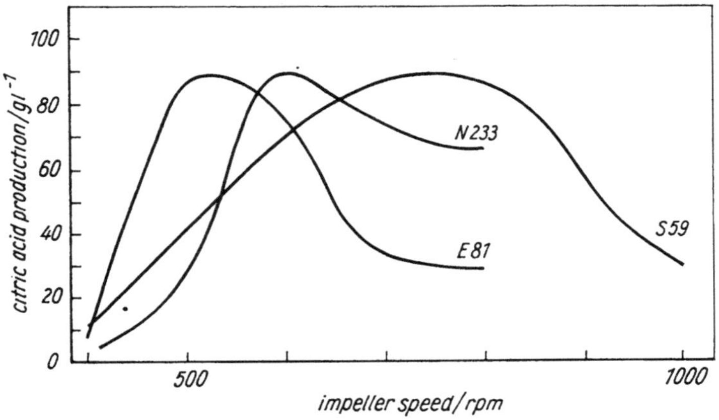

Fig. 2. Dependence of the citric acid production on the impeller speed for different mutants of *Aspergillus niger*

The collision effect is clearly proportional to the circumferential velocity of the impeller blade tips, i.e. to the product dn.:

$$v_{\mathrm{circ}} = \pi d n \approx d n \tag{1}$$

The shear rate D in the tangential jet spreading from the rotating turbine impeller changes linearly with the impeller speed n. We prefer the relation of van't Riet and Smith [6].

$$D \doteq 100 n \tag{2}$$

before the commonly used one proposed by Metzner et al. [3] which is limited to low values of Reynolds mixing number Re_M. The shear stress τ can be than calculated from the experimentally determined flow equation of the given microbial suspension; in our studies the pseudoplastic flow model was used:

$$\tau \doteq 0.8\, D^{0.6} \approx n^{0.6} \tag{3}$$

The relative intensity of turbulence remains almost constant in the vicinity of the impeller and the rootmean-square value of the fluctuating velocity $\sqrt{\overline{v'^2}}$ can be expressed from the data of GÜNKEL and WEBER [2]:

$$\sqrt{\overline{v'^2}} \doteq 0{,}3\pi dn \approx dn \tag{4}$$

It is very difficult to express the cavitation pressure drop Δp quantitavely due to the lack of usable data. The only published ones are those of VAN'T RIET and SMITH [6] from which this equation was derived:

$$\Delta p \doteq 17.4\ \mathrm{Re}_M{}^{0.036} \varrho d^2 n^2 \tag{5}$$

For non-newtonian fluids the apparent viscosity μ_{app} must be used in REYNOLDS mixing number Re_M. By introducing eqn 2 and the relations

$$\mathrm{Re}_M = \varrho d^2 n \mu_{\mathrm{app}}^{-1} \tag{6}$$

$$\mu_{\mathrm{app}} = \tau D^{-1} \doteq 0.8 D^{-0.4} \tag{7}$$

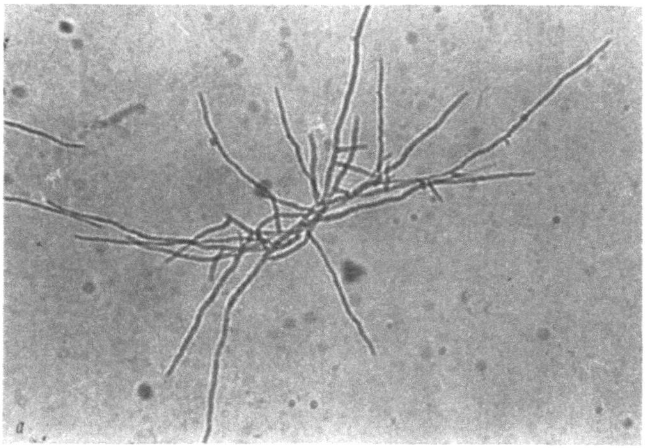

Fig. 3. Morphology of *Aspergillus niger* filaments (magn. 275 ×)
a — impeller speed $n = 400$ rpm

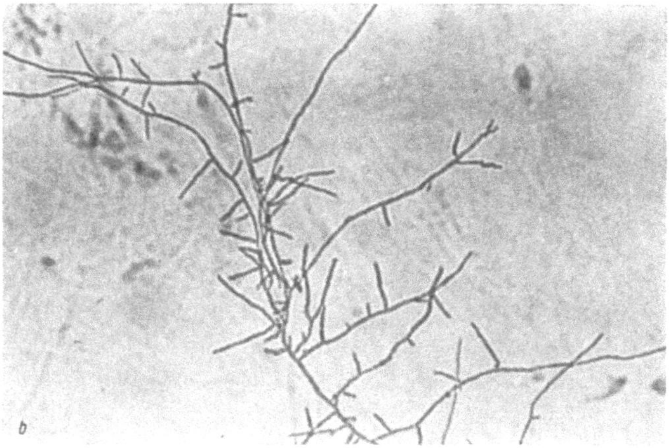

b — impeller speed $n = 1200$ rpm

into eqn 5 this relation is obtained:

$$\Delta p \approx d^{2.07} n^{2.05} \tag{5a}$$

From Fig. 1 follows that the value of circumferential velocity of the impeller blade tips v_{circ} must be kept constant to reach the high production of citric acid. Eqns 4 and 5a show that this condition corresponds with the demands for unchanged root-mean-square value of the fluctuating velocity and for unchanged value of the cavitation pressure drop Δp, while the shear stress effect decreases with the increasing fermenter volume.

To investigate the impeller-induced mechanical effects on microorganisms separately we developed the special air-lift fermenter with controlled circulation of the fermentation medium. The flow pattern is this apparatus can be divided into two regions, one in upward part of the circulation loop where the oxygen is absorbed from rising bubbles, and the other one in non-aerated part of the circulation loop where the mechanical action on microorganisms was performed. The detailed description of the fermenter will be given elsewhere (Biotechnol. Bioengn., in preparation).

The methods for a quantitative and objective appraisal of changes caused by mechanical phenomena are only being developed. The analysis of the mechanical effects on filamentous microorganisms cannot be limited only to the production of the citric acid because the conditions prevailing in the air-lift fermenters fundamentally differ from those in the mechanically agitated ones. The changes in the morphology and physiological state of *Aspergillus niger*, induced by a precisely defined mechanical signal, where quantitavely characterised by the following quantities:

1. lenght of apical parts of hyphae,
2. distance between two neighbouring branching points in intercalary parts of hyphae,
3. cell wall resistance to lytic enzymes from snail gastric juice,
4. chemical composition of the cell wall,
5. change in the amount of the released intracellular nucleotides,
6. changes in the activity of β-glucosidase which mediates the resynthesis of saccharides during the citric acid synthesis, and
7. the citric acid yield.

At the poster section of this Symposium we have presented the poster specificating all the above mentioned methods in detail (see Ujcová et al., "Evaluation of the influence of mechanical forces in agitated fermenter on molds"); their description in this contribution is limited only to a few notes.

The morphological parameters were determined from substantially magnified microphotographs; for every morphological datum the lenghts of 80 hyphae were measured at least.

We have found the monitoring of the released protoplasts to be a very effective method for the description of the cell wall changes. Musílková and Fencl [4, 5] showed that snail gastric juice preferably hydrolyses the apical parts of hyphae from which protoplasts release. Because the mechanical straining of hyphae induces their branching we hoped that the number of the released protoplasts will provide an indirect information on the intensity of branching. However, just the opposite effect was found and the amount of protoplasts released from intensively branched mycelium grown under high mechanical straining was considerably lower. An analysis showed that increasing with the mechanical straining is also the concentration of chitin in the cell wall. We presume that in the mechanically strained filaments chitin is deposed not only in the cell walls in subapical parts of hyphae, but also in the cell walls of apical parts of hyphae in which its concentration is proportional to the straining intensity. Chitin is considerably resistant to the attack of snail lytic enzymes. This fact explains the higher lytic resistance of the cell walls containing an increased amount of chitin.

The quantity of intracellular nucleotides released by the filamentous microorganisms into the fermentation medium changes in dependence on the hydrodynamical conditions in the fermenter. TANAKA et al. [7, 8] used this effect as a criterion for the intensity of mixing in a suspension of filamentous microorganisms. However, UJCOVÁ et al. [9] showed that this criterion cannot be generalised for all species of the filamentous microorganisms. It seems that a change in the amount of released nucleotides characterises changes brought about by the mechanical straining on the permeability of the cell membrane, and not the mass transfer from hyphae to the fermentation medium as proposed by TANAKA et al.

We shall discuss first the response of the *Aspergillus niger* S59 culture to the shear effect. The value of the mean shear rate computed according eqn 2 decreases from $D = 1330\ s^{-1}$ (the laboratory fermenter) to the value $D = 300\ s^{-1}$ (the pilot-scale fermenter with working volume $V = 4.4\ m^3$). The test fermentations were performed at the shear rate $D = 596\ s^{-1}$ (i.e. for shear stress $\tau = 37$ Pa, respectively). The course of one of such an experiment is depicted in Fig. 4. Sixteen hours after the inoculation the

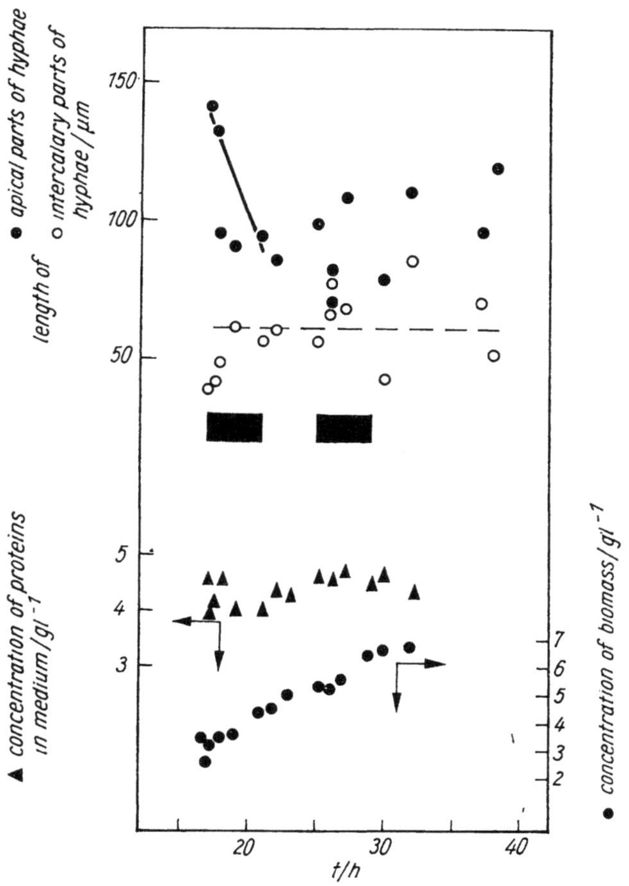

Fig. 4. Morphological changes of *Aspergillus niger* hyphae resulting from shear stress in the liquid phase (black rectangles denote the duration of the mechanical effect)

cultivation proceeded in the alternating switching on and off the spinning of shear stress generator in four-hour intervals (the black rectangles in Fig. 4 represent the spinning period); the four-hour interval corresponds approximately with the population's generation time, i.e. with the time necessary for the biomass doubling. This pattern of cultivation rendered possible analysis of the new filaments grown under the defined hydrodynamics conditions at the end of every four-hour interval. The intactness of hyphae was assessed from microscopic observations and from monitoring the concentration of proteins in the fermentation medium. Switching on the shear stress generator brought about a reduction of the apical hyphae lenght to approximately one half. Further cultivation with alternating switching on and off did not induce a significant change in the morphology. The initial decrease in the lenght of the apical hyphae is probably connected with the culture's growth mode; the mould first grows in pellets and after the first action of shear stress generator starts to grow diffusively.
Let us concentrate now on the effect of the impeller blades collision with the mould filaments. An explicit dependence of the lenght of the apical hyphae on switching on and off the collision generator was found for the circumferential velocity $v_{circ} = 6.3\ \mathrm{ms^{-1}}$ (Fig. 5); similar values of v_{circ} are typical for many industrial fermentations (Einsele, A.: "Scaling-up of bioreactors. Theory and reality.", communication at the V. International

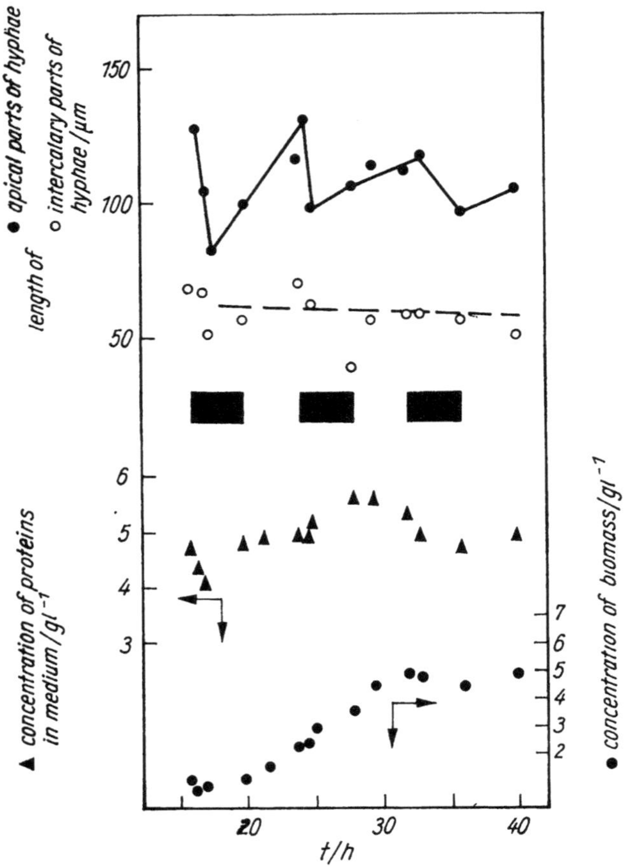

Fig. 5. Morphological changes of *Aspergillus niger* hyphae resulting from collision effect (black rectangles denote the duration of the mechanical effect)

Fermentation Symposium, West Berlin (1976)), but this value is higher than that one which was found to be optimal for our process. The course of the cultivation was similar to the shear stress experiment. The lenght of the apical hyphae again decreased to one half after the first switching on the collision generator. The same effect was observed after the second period of the generator action. After the third period the effect was less profound, probably because of the limitation by nutrients and consequently slower growth rate of the mould.

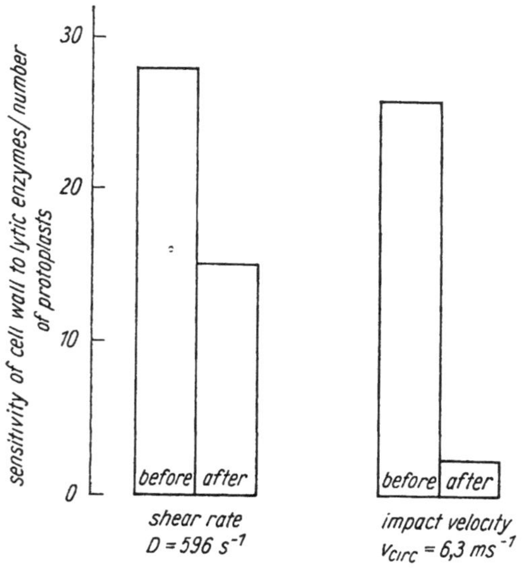

Fig. 6. Change of the lytic enzymes resistance caused by mechanical influence of shear stress and collision

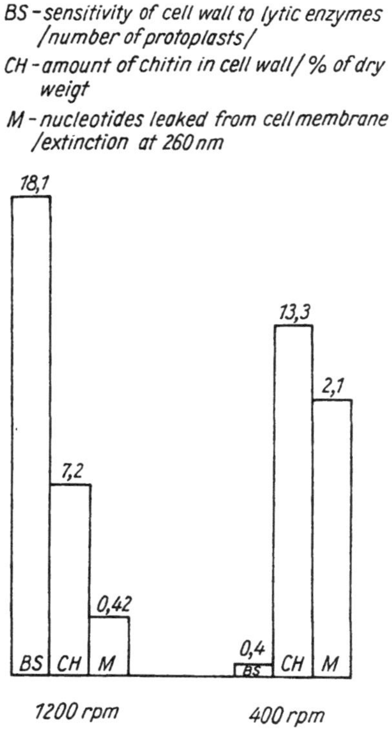

Fig. 7. Changes in the properties of the cell wall and membrane induced by different impeller speed

The importance of the mechanical effects on the cell wall is evident from Fig. 6. The amount of released protoplasts before and after mechanical straining significantly differs, especially in the case of collision effect. We therefore presume that the shear stress does not represent a decisive factor in the group of the impeller-induced mechanical effects on the filamentous microorganisms.

The changes in the physiological state of *Aspergillus niger* S59 culture during the cultivation in the laboratory fermenter at two limiting impeller speeds are given in **Fig. 7.** With the increasing impeller speed the lenght of apical hyphae decreases. The changing impeller speed also significantly affects the activity of β-glucosidase, the production of citric acid (Fig. 8), the resistance of the cell wall to the effect of lytic enzymes (this resistance being dependent on the change of the chemical composition of the cell wall), and the permeability of the cell membrane.

Van't Riet found that with the constant value of the circumferential velocity v_{circ} the cavitational pressure drop behind the impeller blade Δp slightly increases with the impeller diameter (with doubling of the impeller diameter the pressure-drop increases by several percents — see eqn 5). We cannot exclude the influence of cavitation on the physiological state of microorganisms, but in our process the value of Δp remains constant during scale-up and, therefore, does not change the biosynthetic activity of *Aspergillus niger*. However, the separated effect of cavitation on the morphology and the physiological state we have not yet investigated experimentally.

Our work was prompted by the lack of objective criteria for the scale-up of the submerged citric acid production. We found the decisive effect of the direct impact of the impeller-blades on the mould filaments which implies the necessity of a constant circumferential velocity of the impeller blade tips. This would ensure a defined branching and septating of hyphae and thus also the required physiological state of the filamentous microorganism such as the differentiation along the hypha. This in turn ensures optimal

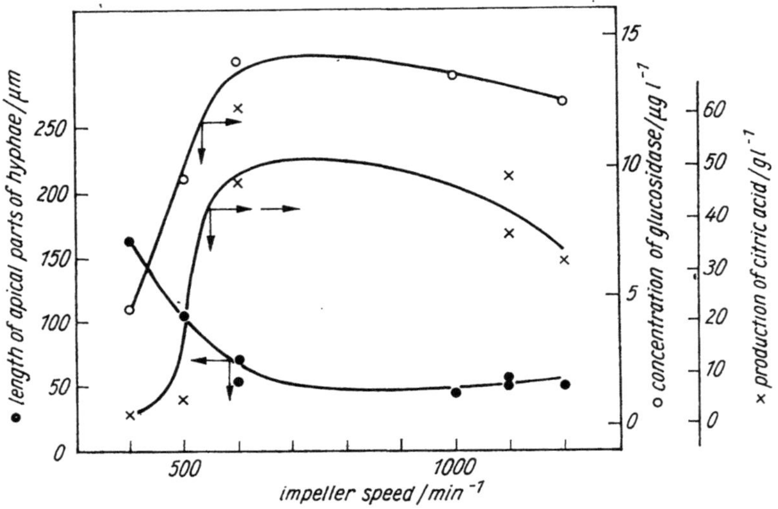

Fig. 8. Changes of morphology, β-glucosidase activity and citric acid production induced by impeller speed

conditions for the production of citric acid in the fermenter. We found that for the maximal production of citric acid the studied production mutants of *Aspergillus niger* require specific cultivation conditions. Our results cannot be therefore generalised and for the every individual process the optimal chemical, physico-chemical and hydrodynamical conditions have to be found.

While there exist theories on the effects of chemical and physico-chemical conditions on the enzyme activities in microorganisms, there is no explanation for the observed mechanical phenomena. The existing hypotheses give no explicit answers. On the other hand the classification of the impeller-induced effects on microorganisms into the separate categories renders possible a better understanding of conditions for optimization of the fermentation process and thus make possible its scale-up on scientific base.

Nomenclature

D	shear rate	s^{-1}
d	impeller diameter	m
n	impeller speed	s^{-1}

P	impeller power consumption	W
Δp	cavitation pressure drop	Pa
Re_M	Reynolds mixing number	1
V	fermenter working volume	m^3
v_{circ}	circumferential velocity of the impeller blade tips	ms^{-1}
$\sqrt{\overline{v'^2}}$	root-mean-square value of the fluctuating velocity	ms^{-1}
μ_{app}	apparent viscosity	Pas
ϱ	density	kgm^{-3}
τ	shear stress	Pa

References

[1] Dion, W. M., Carilli, A., Sermonti, G. and Chain, E. B.: The effect of mechanical agitation on the morphology of *Penicillium chrysogenum* in stirred fermenters; Rend. Ist. Super. Sanita, English Ed. **17** (1954) 187.

[2] Günkel, A. A. and Weber, M. E.: Flow phenomena in stirred tanks; AIChE J. **21** (1975) 931.

[3] Metzner, A. B., Feehs, R. H., Ramos, H. L., Otto, R. E., and Tuthill, J. D.: Agitation of viscous newtonian and non-newtonian fluids; AIChE J. **7** (1961) 3.

[4] Musílková, M. and Fencl, Z.: Formation of protoplasts in *Aspergillus niger*; Folia microbiologica **11** (1966) 472.

[5] Musílková M. and Fencl, Z.: Some factors affecting the formation of protoplasts in *Aspergillus niger*; Folia microbiologica **12** (1968) 235.

[6] Van't Riet, K. and Smith, J. M.: The trailing vortex system produced by Rushton turbine agitators; Chem. Engng. Sci. **30** (1975) 1093.

[7] Tanaka, H., Takahashi, J. and Ueda, K.: A standard for the intensity of agitation shock on mycelia on agitation of mycelial suspensions; J. Ferment. Technol. (Japanese) **53** (1975) 18.

[8] Tanaka, H. and Ueda, K.: Kinetics of mycelial growth accompanied by leakage of intracellular nucleotides caused by agitation; J. Ferment. Technol. (Japanese) **53** (1975) 27.

[9] Ujcová, E., Fencl, Z., Musílková, M. and Seichert, L.: Dependence of release of nucleotides from fungi on fermentor turbine speed; Biotechnol. Bioengn. **22** (1980) 237.

BOOK REVIEW

W. W. Kafarow, A. J. Winarow, L. S. Gardejew

Modellierung biochemischer Reaktoren

Lesnaja Promyshlenost, Moskau 1979 (russ.).
342 S., 125 Abbildungen und 19 Tafeln, Leinen, 3,60 Rbl

Die Biotechnologie als Technologie biologischer Stoffwandlungsprozesse befindet sich bekanntermaßen in einer raschen Entwicklung. Wie erfolgreich diese Entwicklung weitergehen wird, hängt auch davon ab, wie es gelingt, das Wissen der beteiligten naturwissenschaftlichen, technischen, ökonomischen und gesellschaftswissenschaftlichen Disziplinen auf neue Prozesse anzuwenden und in ökonomisch effektive technische Verfahren umzusetzen.
Daraus ergeben sich an die beteiligten Bearbeiter hohe, z. T. ungewohnte Anforderungen an ein technologisch orientiertes Denken. Die bisher vorhandene Literatur auf dem Gebiet der Biotechnologie trägt dieser Forderung kaum Rechnung.
Es ist deshalb erfreulich, daß mit dem vorgestellten Titel ein Buch vorgelegt wird, dessen wesentliches Merkmal gerade die durchgängige Anwendung dieser Denkweise in der Biotechnologie ist. Das war freilich bei einem Autorenkollektiv unter Mitwirkung eines Verfahrenstechnikers vom Range Kafarows auch zu erwarten.
Das Buch gliedert sich in 6 Hauptabschnitte:

- Systemanalyse als Methode zur Untersuchung von biotechnischen Produktionsprozessen
- Modellierung der mikrobiellen Wachstumskinetik in biochemischen Reaktoren
- Hydrodynamik biochemischer Reaktoren
- Stofftransport in biochemischen Reaktoren
- Modelle biochemischer Reaktoren
- Prozeßanalyse und Optimierung biotechnologischer Systeme

Den Kapiteln sind jeweils Zusammenstellungen von Literatur beigefügt.
Es zeigt sich, daß eine große Zahl von Ergebnissen aus Originalergebnissen der Autoren in das Buch aufgenommen worden sind, die auf diese Weise dem Leser bequem zur Verfügung stehen. Sicher macht dieser Umstand — insbesondere für den mit der technologischen Denkweise bereits vertrauten Leser — einen wesentlichen Teil des Wertes des Buches aus.
Die Bedeutung der einzelnen Kapitel ist unterschiedlich. Den Fragen der Hydrodynamik und des Stoffüberganges (Kap. 3 und 4) wird breiterer Raum gegeben. Hier werden neben allgemeinen theoretischen Gesichtspunkte auch Modelle und Modellkonstanten spezieller Stoffsysteme (z. B. für die Biomasseproduktion auf unkonventionellen Substraten) angegeben. Technische Realisierungen von Fermentoren werden im Kapitel 5 naturgemäß nur kurz angesprochen und spezielle Modelle nicht angegeben. Auf die Anwendbarkeit bestimmter Fermentoren auf spezielle Prozesse wird hingewiesen.
Die Ausführungen zum Komplex der Prozeßanalyse und der Optimierung biotechnischer Systeme (Kapitel 6) sind eher geeignet, als grundlegende Information über die Methodik der Optimierung zu dienen. Sie werden dem Nichtfachmann helfen, eine zweckmäßige Zusammenarbeit mit Spezialisten zu ermöglichen und Zugang zu den in der DDR veröffentlichten weiterführenden Arbeiten, insbesondere auch Kafarows, zu diesem Komplex zu finden.
Insgesamt ist das Buch eine wertvolle Bereicherung der biotechnologischen Literatur. Neben seiner Bedeutung für den Praktiker ist es — nicht zuletzt wegen seiner guten Ausstattung und vielen instruktiven Bildern — auch für die Ausbildung zu empfehlen.

K. Soyez

Acta Biotechnologica 1 (1981) 4, 339–350

Bioconversion of Lipophilic Compounds by Immobilized Microbial Cells in Organic Solvents

S. FUKUI and A. TANAKA

Laboratory of Industrial Biochemistry, Department of Industrial Chemistry, Faculty of Engineering, Kyoto University, Yoshida, Sakyo-ku, Kyoto 606, Japan

Summary

Microbial cells were gel-entrapped with photo-crosslinkable resin prepolymers or urethane prepolymers, respectively. The resulting gels have different tailor-made hydrophobic or hydrophilic character. They were used for successful bioconversion of hydrophobic steroids and terpenoids in water-saturated mixtures of organic solvents. The experiments show the influence of the hydrophobicity of the gels and the polarity of the solvent mixtures, respectively. Use of hydrophobic gels and less polar solvents is preferable for bioconversion of hydrophobic compounds. The selective formation of a desired product among diverse products from a single substrate by appropriate use of hydrophobic or hydrophilic gels is possible. In each case, tests should be made to select the appropriate gel and solvent mixture. Bioconversions tested are: dehydroepiandrosterone to 4-androstene-3,17-dione; cholesterol to cholestenone; β-sitosterol to β-sitostenone; stigmasterol to stigmastenone; pregnenolone to progesterone; testosterone to Δ^1-dehydrotestosterone or 4-androstene-3,17-dione, respectively; all with immobilized cells of *Nocardia rhodocrous*; and stereoselective hydrolysis of dl-menthyl-succinate to yield l-menthol with immobilized cells of *Rhodotorula minuta* var. *texensis*.

Zusammenfassung

Mikroorganismenzellen wurden durch Geleinschluß immobilisiert, wobei photosensitive Harz- bzw. Urethanpräpolymere als Ausgangsstoffe für die Vernetzung dienten. Die entstandenen Gele besitzen „maßgeschneiderte" hydrophobe bzw. hydrophile Eigenschaften und wurden erfolgreich zur Biokonversion von hydrophoben Steroiden und Terpenoiden in wassergesättigten Gemischen organischer Lösungsmittel verwendet. Die Experimente zeigen den Einfluß der Hydrophobie der Gele bzw. der Polarität der Lösungsmittelgemische. Der Einsatz hydrophober Gele und weniger polarer Lösungsmittel ist für die Biokonversion von hydrophoben Verbindungen günstiger. Die selektive Bildung eines gewünschten Produktes bei einer möglichen Vielzahl von Produkten aus einem einzigen Substrat ist durch geeignete Verwendung hydrophober und hydrophiler Gele möglich. Dabei sind stets Tests durchzuführen, um das geeignete Gel und das Lösungsmittelgemisch auszuwählen. Folgende Transformationen wurden untersucht:

Dehydro-epi-Androsteron	→	4-Androsten-3,17-dion
Cholesterol	→	Cholestenon
β-Sitosterol	→	β-Sitostenon
Stigmasterol	→	Stigmastenon
Pregnenolon	→	Pregnesteron
Testosteron	→	Δ^1-Dehydro-Testosteron
Testosteron	→	4-Androsten-3,17-dion

Für die aufgeführten Transformationen wurden immobilisierte Zellen von *Nocardia rhodocrous* verwendet; für die stereoselektive Hydrolyse von dl-Menthyl-Succinat zu l-Menthol wurden immobilisierte Zellen von *Rhodotorula minuta* var. *texensis* eingesetzt.

Introduction

In the case of bioconversion of highly lipophilic compounds, it will be desirable to carry out enzymatic reactions in a mixture of water and a suitable organic cosolvent or in an appropriate organic solvent system, if the catalytic activity of enzyme is maintained in such a reaction system. The use of organic solvent will improve the poor solubility in water of the substrate or other reaction components of hydrophobic nature, and shift an unfavourable thermodynamic equilibrium to a desired direction if necessary.

However, attempts to make free, native enzyme function in an organic solvent system have not been successful. Usually, catalytic activity of enzyme drastically decreases and its substrate specificity disappears. One of the most probable inactivation mechanism of enzyme would be distortion of catalytically active conformation of enzyme molecule.

In general, immobilization is one of the most promising approaches for enzyme stabilization. The conformational structure of enzyme molecule immobilized on or in a rigid support will become more resistant against distortion, and thus, unfolding of the high dimensional structure of enzyme which leads to loss of catalytic activity will become much more difficult than the case of free enzyme. One of successful application of carrier-bound enzyme to reaction in organic solvent is the production of N-acetyl-l-tryptophan ethyl ester from N-acetyl-l-tryptophan and ethanol in chloroform, as reported by KLIBANOV *et al.* [1].

It is well known that, *in vivo*, many enzymes, especially those catalyzing transformation of lipophilic compounds, function in membrane-bound state, and that the stability of such membrane-bound enzyme is in general much greater than that of the enzyme released from membrane. In fact, enzymes contained in microbial cells are, in some cases, useful for *in vitro* bioconversion of lipophilic compounds in organic solvent systems. For instance, BUCKLAND *et al.* [2] have reported the conversion of cholesterol to cholestenone by the cells of *Nocardia* sp. in water-immiscible organic solvents. However, the operational stability of free cells seems not satisfactory.

Inclusion of enzymes within suitable gels would give microenvironment analogous to that of membrane-bound enzymes. Multi-point interactions between entrapped enzymes and gel matrices would give stabilizing effect. Thus, one can expect that entrapment of microbial cells in suitable gels will render the enzymes in cells more stabilized state. Even if membrane-bound enzymes are liberated from membrane, these enzymes will be kept within gel matrices.

KLIBANOV [3] has asserted that, if biocatalysts are attached to or included in suitable material which has effects to lower the concentration of organic solvent in the environment of enzymes or exclude organic solvent from the vicinity of enzymes, stabilization of so immobilized biocatalysts will be brought about. Such the effect could be obtained by the use of gel material of hydrophilic nature.

However, the situation of immobilized biocatalysts to be used for conversion of lipophilic compounds in organic solvent system is more complicated. Affinity of lipophilic substrate for gels entrapping biocatalysts and diffusion of reactants through gel matrices are important factors. Low affinity of hydrophilic gels for hydrophobic substrates will lower the apparent activity of the gel-entrapped biocatalysts. Thus, use of suitable hydrophobic gels should be preferable depending on hydrophobicity of substrates and polarity of solvents to be used.

In this monograph we would like to summarize our experimental results on the bioconversion of lipophilic compounds by immobilized microbial cells in organic solvents, emphasizing the effect of gel hydrophobicity. Effects of polarity of solvents and hydrophobicity of substrates on such reactions are also described. However, the bioconversion in water and water-miscible cosolvent systems, such as dehydrogenation of hydrocortisone to prednisolone [4, 5], is not included in this article.

Entrapment of microbial cells

We have developed novel, convenient methods to entrap biocatalysts inside gel matrices formed from synthetic prepolymers [6]. Figures 1 and 2 show the structures of the prepolymers of photo-crosslinkable resin (ENT and ENTP) and the prepolymers of ur-

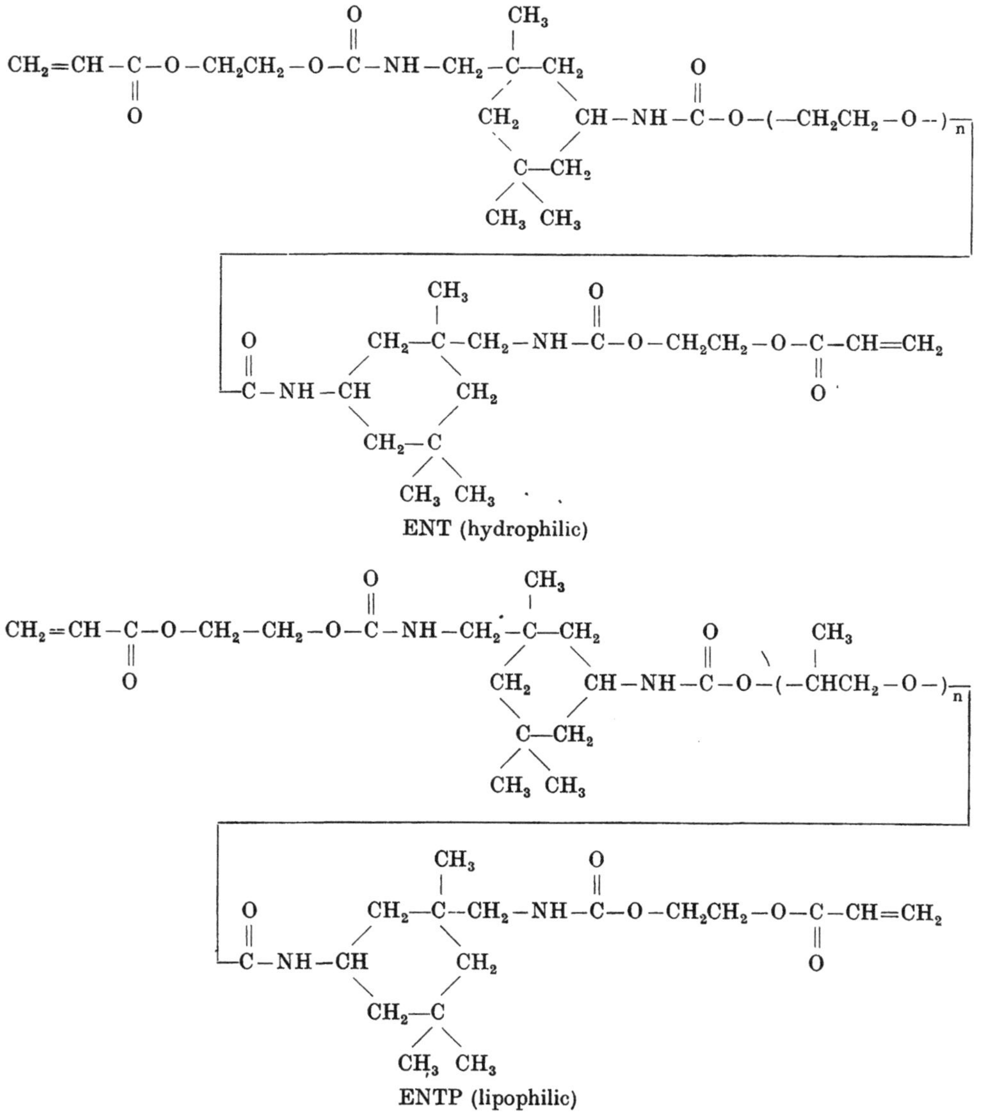

Fig. 1. Structures of typical photo-crosslinkable resin prepolymers. ENT is water-soluble and gives hydrophilic gels. ENTP is water-insoluble and gives hydrophobic gels.

ethane resin (PU), respectively, which have been used in our experiments. Photo-cross-linkable resin prepolymers have photo-sensitive functional groups, such as acryloyl group, at the terminals of linear chain. The chain length of prepolymers can be adjusted by using poly(ethylene glycol) or poly(propylene glycol) of optional chain length in the linear chain [7]. Thus, ENT-4000 means that the prepolymer contains poly(ethylene glycol)-4000 (average Mw, about 4,000; chain length, about 40 nm) as main skeleton. When the main skeleton consists of poly(ethylene glycol), the prepolymer and accordingly, the gels formed from the prepolymer, should have a hydrophilic character. On the other hand, when poly(propylene glycol) composes the main skeleton, the prepolymer and thus, the gels to be formed, should be hydrophobic. ENT-4000 gives hydrophilic gels, while ENTP-2000 hydrophobic gels. In the case of urethane prepolymers which have isocyanate groups at both terminals of linear chain, prepolymers with different hydrophilic or hydrophobic character can be obtained by changing the ratio of the poly(ethylene glycol) part and poly(propylene glycol) part in the polyether moiety.

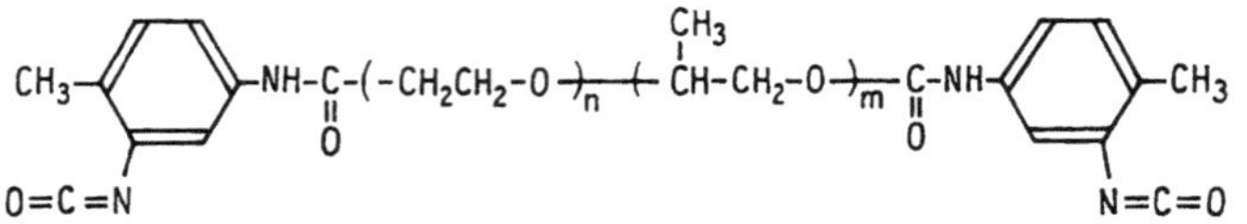

PU prepolymer

Prepolymer	*Mw of polyol*	*NCO content*	*Ethylene glycol content*
PU-3	*2529*	*4.2 %*	*57 %*
PU-6	*2627*	*4.0 %*	*91 %*
PU-9	*2616*	*4.0 %*	*100 %*

Fig. 2. Structure and properties of typical urethane prepolymers.
PU-3, PU-6 and PU-9 are water-miscible. PU-3 gives hydrophobic gels, while PU-6 and PU-9 hydrophilic gels.

PU-3 with a high content of poly(propylene glycol) gives hydrophobic gels, while PU-6 and PU-9 with a high content of poly(ethylene glycol) hydrophilic gels. The chain length of the prepolymers also can be changed [5]. Furthermore, incorporation of ionic group, cationic or anionic group, into the main skeleton gives the gels optionally ionic nature [6].

Gelation of photo-crosslinkable resin prepolymers can be easily completed by illuminating the mixture of prepolymer solution, a small amount of photo-sensitizer, such as benzoin ethyl ether or benzoin isobutyl ether, and enzyme solution or cell or organelle suspension by near-ultraviolett ray for 3—5 min [8—10]. Entrapment of biocatalysts with urethane prepolymers is much simpler. When liquid prepolymers are mixed with aqueous solution of enzymes or aqueous suspension of organelles or cells, prepolymers react each other, being crosslinked by forming urea linkage with liberation of carbon dioxide [11, 12].

Specific features of the prepolymer methods mentioned above are not only their simplicity in entrapping processes but also tailor-made network of gels with desired physical and chemical properties can be obtained by the use of appropriate prepolymers. Another merit is that monomers, such as acrylamide etc., which are liable to give bad effects on the activity of biocatalysts during the immobilization processes and, if remain in gels, are undesirable from sanitary viewpoints, are not involved.

Bioconversion of steroids

Transformation of steroids is one of the most interesting subjects in bioconversion of hydrophobic substrates by biocatalysts entrapped in suitable gels. As seen in Figure 3, *Nocardia rhodocrous* mediates different types of bioconversion of a variety of steroids; Δ^1-dehydrogenation, 17β-hydroxysteroid dehydrogenation and 3β-hydroxysteroid dehydrogenation [13–17]. In order to study the effect of gel hydrophobicity, that is, the influence of affinity between lipophilic substrate and gels entrapping microbial cells, a very simple parameter, the partition coefficient of substrate between gels and external solvent, was employed.

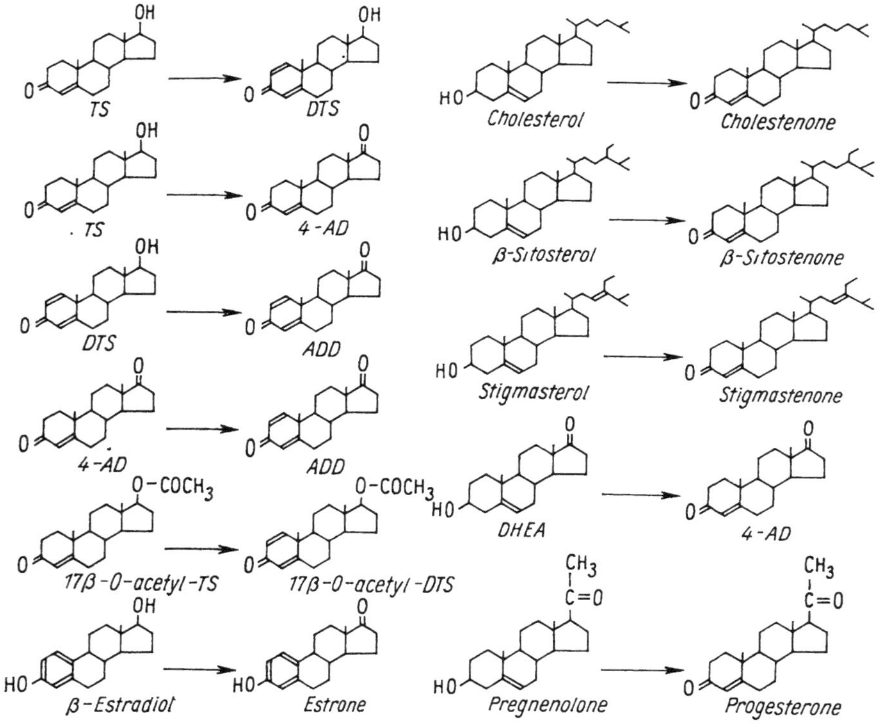

Fig. 3. Steroid transformation catalyzed by free and entrapped *Nocardia rhodocrous* cells in organic solvents.
Abbreviation used: 4-AD, 4-androstene-3,17-dione; ADD, 1,4-androstadiene-3,17-dione; DHEA, dehydroepiandrosterone; DTS, Δ^1-dehydrotestosterone; TS, testosterone.

The effect of gel hydrophobicity was investigated on the conversion of 3β-hydroxy-Δ^5-steroids into 3-keto-Δ^4-steroids in water-saturated mixture of benzene and *n*-heptane (1 : 1 by volume) as solvent [14]. The solvent system was chosen based on the following criteria: Substrate and product are enough soluble; enzyme activity is not damaged; solvent does not render the gels swollen. As shown in Table 1, *N. rhodocrous* cells entrapped in hydrophobic gels, such as ENTP-2000 and PU-3, showed the activity to convert dehydroepiandrosterone (DHEA) into 4-androstene-3,17-dione (4-AD) comparable to that of the free cells. On the other hand, the cells entrapped in hydrophilic gels, ENT-4000 and PU-6, were less active. Figure 4B shows the relationship between the

Table 1. Effect of Gels on Steroid Transformation by *Nocardia rhodocrous* in Non-Polar Solvent

Cell	Relative activity (%) on			
	Cholesterol	β-Sitosterol	Stigmasterol	DHEA
Free	100	100	100	100
ENTP-2000-entrapped	102	83	111	107
ENT-4000-entrapped	0	0	0	78
PU-3-entrapped	77	54	85	105
PU-6-entrapped	0	—	0	72
PU-9-entrapped	0	0	—	—

DHEA, Dehydroepiandrosterone
Reaction solvent, Water-saturated benzene-*n*-heptane (1 : 1 by volume)

relative activity of entrapped cells and the partition coefficient of DHEA between the gels and external solvent, namely, affinity of substrate for gels entrapping cells, both of which changed depending on the hydrophobicity of gels. The abscissa shows the mixing ratio of hydrophobic prepolymer PU-3 and hydrophilic prepolymer PU-6. A close relationship between the relative activity of the gel-entrapped cells and the partition coefficient of substrate can be observed. The half-life of the free cells in the conversion of DHEA in benzene-*n*-heptane (1 : 1 by volume) was calculated to be 9—10 hr, while that of the entrapped cells was estimated to be at least 20—24 hr, indicating the improved stability of the immobilized cells.

In the case of transformation of cholesterol, β-sitosterol and stigmasterol, more lipophilic substrates, to the corresponding 3-keto-Δ^4-steroids, much more clear-cut interrelationship was observed between gel hydrophobicity and activity of gel-entrapped cells (Table 1).

Only the cells entrapped with hydrophobic prepolymers (ENTP-2000 and PU-3) exhibited the catalytic activity. This phenomenon was also confirmed when the hydro-

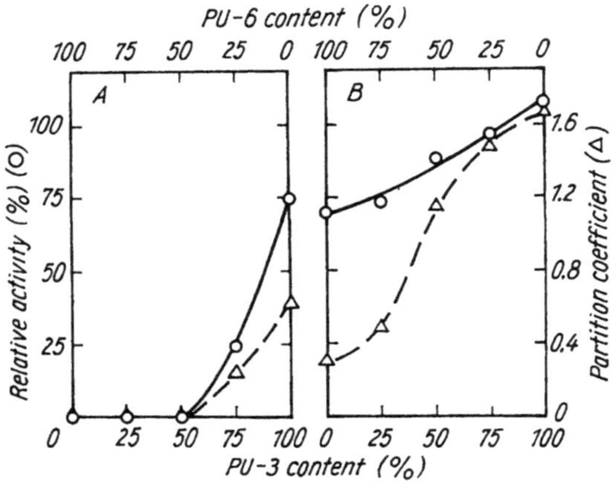

Fig. 4. Effect of hydrophobicity of PU resins on steroid transformation activity of *Nocardia rhodocrous* and partition coefficient in non-polar solvent systm
A, cholesterol; B, dehydroepiandrosterone. Reaction solvent, water-saturated benzene-*n*-heptane (1 : 1 by volume). (○), Relative activity compared with that of free cells; (△), partition coefficient.

phobicity of gels was changed by mixing PU-3 and PU-6 (Fig. 4A). No activity of cholesterol conversion, in accordance with the low partition coefficient of cholesterol, was observed at the lower ratio of PU-3, a hydrophobic prepolymer.

In addition to gel hydrophobicity and substrate hydrophobicity, polarity of reaction solvent also gave a marked effect on the conversion of steroids [15]. Although *N. rhodocrous* cells entrapped in hydrophilic gels could not transform cholesterol in non-polar solvent, benzene-*n*-heptane (1 : 1 by volume), the substitution of benzene with chloroform made the hydrophilic gel-entrapped cells active with the increased partition coefficient of cholesterol (Fig. 5). However, enhancement of solvent polarity fairly lowered the activity and stability of the free cells and subsequently, those of the entrapped cells. In water-saturated mixture of chloroform and *n*-heptane (1 : 1 by volume), little effect of

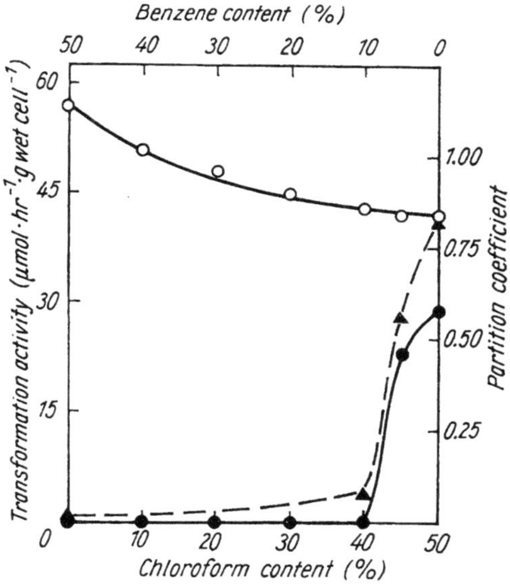

Fig. 5. Effect of solvent polarity on cholesterol transformation by free and ENT-4000-entrapped *Nocardia rhodocrous* cells.
Reaction solvents were prepared by mixing benzene and chloroform in different ratios keeping the content of *n*-heptane constant (50% by volume). (○), Activity of free cells; (●), activity of ENT-4000-entrapped cells; (▲), partition coefficient of cholesterol.

gel hydrophobicity was observed on the conversion of cholesterol and DHEA (Fig. 6), differing from the results obtained in the non-polar solvent system shown in Figure 4. The hydrophobic gel-entrapped cells rather showed a low activity in the conversion of pregnenolone, a less hydrophobic substrate which was not soluble in benzene-*n*-heptane (Fig. 6). In spite of these facts, usefulness of hydrophobic gels is clear in the transformation of highly hydrophobic compounds because the transformation activity of the gel-entrapped cells is usually high and stable in less polar solvents. In conclusion, it would be very important for successful bioconversion of lipophilic compounds to select gel material with suitable hydrophobicity and reaction solvents with suitable polarity, both of which should be determined depending on substrate hydrophobicity.

The next example is the effect of gel hydrophobicity on conversion routes from testosterone (TS) to 1,4-androstadiene-3,17-dione (ADD) [17]. Figure 7 illustrates the transformation pathway of TS into ADD mediated by *N. rhodocrous* in water-saturated benzene-n-heptane (4 : 1 by volume) mixture. In the presence of an electron acceptor, such as phenazine methosulfate (PMS), the free bacterial cells converted TS to ADD *via* two diverse routes. As shown in the figure, 17β-dehydrogenation product, 4-androstene-3,17-dione (4-AD), and Δ^1-dehydrogenation product, Δ^1-dehydrotestosterone (DTS), appeared as the intermediates. In these reactions, Δ^1-dehydrogenation required PMS absolutely, whereas 17β-dehydrogenation could proceed without exogenous electron acceptor, although PMS stimulated the reaction. When the cells were entrapped in gels of different hydrophobicity or hydrophilicity, the property of gels gave striking effects on the conversion routes. With hydrophobic gel (PU-3-rich)-entrapped cells, 4-AD was formed as major reaction product. On the other hand, DTS was the main product with hydrophilic gel (PU-6-rich)-entrapped cells. This different profile in dehydrogenation

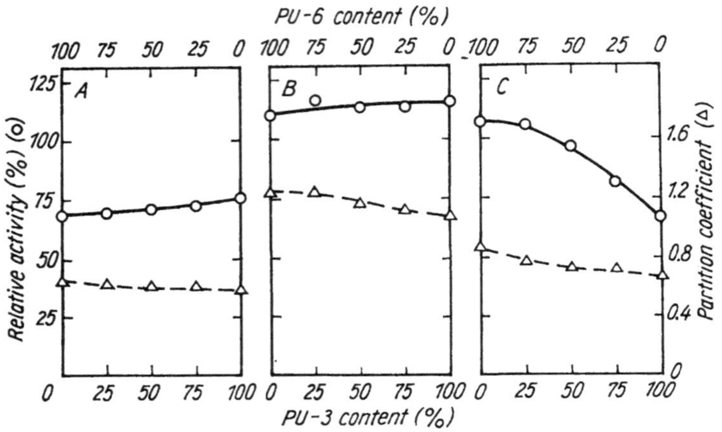

Fig. 6. Effect of hydrophobicity of PU resins on steroid transformation activity of *Nocardia rhodocrous* and partition coefficient in polar solvent system.
A, cholesterol; B, dehydroepiandrosterone; C, pregnenolone. Reaction solvent, water-saturated chloroform-n-heptane (1:1 by volume). (○), Relative activity compared with that of free cells; (△), partition coefficient.

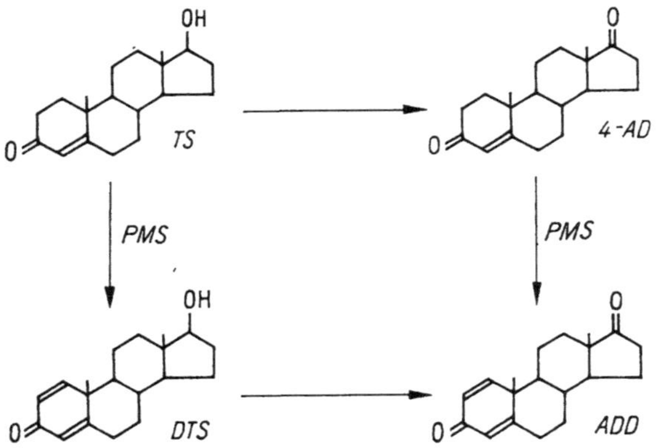

Fig. 7. Transformation of testosterone catalyzed by *Nocardia rhodocrous*. Abbreviation used: 4-AD, 4-androstene-3,17-dione; ADD, 1,4-androstadiene-3,17-dione; DTS, Δ^1-dehydrotestosterone; TS, testosterone; PMS, phenazine methosulfate.

products can be explained by a marked difference in the affinity of PMS, a hydrophilic compound, for hydrophobic gels and hydrophilic ones. With hydrophilic gel-entrapped cells, Δ^1-dehydrogenation of TS to yield DTS is stimulated by PMS uptaken inside gels and DTS so formed inhibits 17β-hydroxysteroid dehydrogenase converting DTS to ADD. On the contrary, in the case of cells entrapped in hydrophobic gels hardly uptaking PMS, 17β-dehydrogenation of TS becomes the main route.
Thus, the hydrophilic or hydrophobic nature of the gels was found to be an important factor for controlling the product formation. The selective formation of a desired product among diverse products from a single substrate by the appropriate use of hydrophilic or hydrophobic gels to entrap cells would be applicable to the bioconversion of many organic compounds.

Stereoselective hydrolysis of *dl*-menthyl ester

The last example is the stereoselective hydrolysis of *dl*-menthyl succinate by gel-entrapped cells of *Rhodotorula minuta* var. *texensis* in water-saturated *n*-heptane (Fig. 8) [18]. *l*-Menthol, which has a peppermint flavour, can be obtained by stereospecific hydrolysis of an appropriate ester of chemically synthesized *dl*-menthol. Chemically synthesized *dl*-menthol contains four isomers, *dl*-menthol, *dl*-iso-menthol, *dl*-neo-menthol and *dl*-iso-neo-menthol, and separation of *l*-menthol from the isomers and racemates is industrially important.

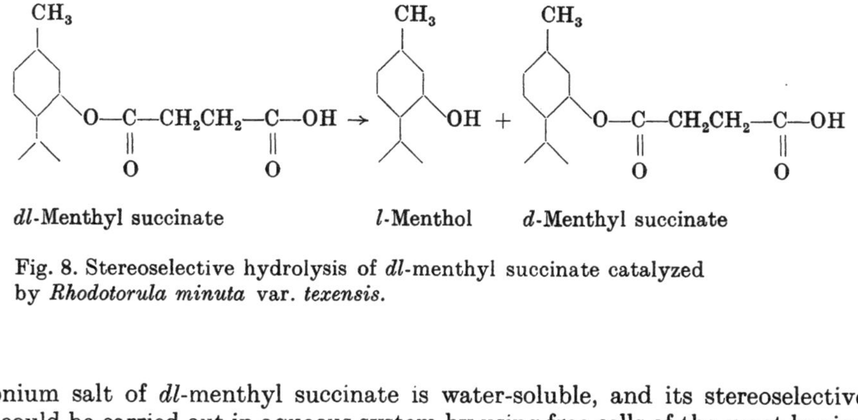

dl-Menthyl succinate *l*-Menthol *d*-Menthyl succinate

Fig. 8. Stereoselective hydrolysis of *dl*-menthyl succinate catalyzed by *Rhodotorula minuta* var. *texensis*.

The ammonium salt of *dl*-menthyl succinate is water-soluble, and its stereoselective hydrolysis could be carried out in aqueous system by using free cells of the yeast having the esterase activity. However, most part of *l*-menthol formed was not dissolved in aqueous buffer, and accumulated on the surface of yeast cells, thus decreasing the activity. To prevent the accumulation of the product, *l*-menthol, various kinds of water-miscible organic solvents were tested as cosolvent. As shown in Table 2, the hydrolytic activity of the yeast cells was reduced significantly in the presence of organic cosolvents. Hydrolysis of menthyl succinate by the free cells in organic solvents was very difficult because the cells could not be suspended well in such solvents. Among the two-phase systems consisting of potassium phosphate buffer and organic solvents, *n*-heptane was found to be most suitable. Thus, in the case of the gel-entrapped cells, also, water-saturated *n*-heptane was employed to obtain a homogeneous reaction system. Although the effect of gel hydrophobicity was not so remarkable as the case of steroids, the activity of gel-entrapped cells increased along with increased gel hydrophobicity. Figure 9 shows the comparison of operational stability between free cells and PU-3-entrapped

Table 2. Effect of Solvents on Stereoselective Hydrolysis of *dl*-Menthyl Succinate by Free Cells of *Rhodotorula minuta* var. *texensis*

Substrate	Solvent	Relative activity
dl-Menthyl succinate ammonium salt	KPB[a]	100
	KPB : Methanol (75 : 25)	67
	KPB : Dimethylformamide (50 : 50)	19
	KPB : CH_3CN (50 : 50)	7
	KPB : CH_3CN (80 : 20)	42
dl-Menthyl succinate	Water-saturated benzene	1
	Water-saturated chloroform	0
	Two-phase system (KPB : Organic solvent, 1 : 5)	
	t-Butyl acetate	0
	Benzene	32
	Benzene-*n*-Heptane (50 : 50)	35
	Benzene-*n*-Heptane (30 : 70)	53
	Benzene-*n*-Heptane (10 : 90)	75
	n-Heptane	102

[a] KPB, 20 mM Potassium phosphate buffer (pH 7.0)

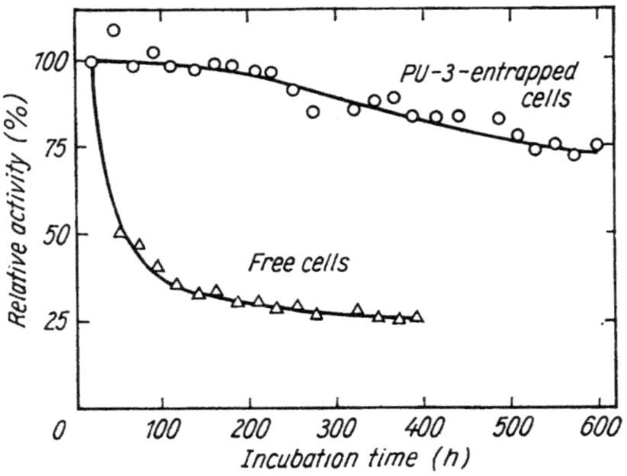

Fig. 9. Repeated use of free and PU-3-entrapped *Rhodotorula minuta* var. *texensis* in stereoselective hydrolysis of *dl*-menthyl succinate.
Each reaction was carried out for 24 hr in buffer-*n*-heptane two-phase system (1:5 by volume) (free cells) or in water-saturated *n*-heptane (entrapped cells). (○), PU-3-entrapped cells; (△), free cells.

cells over repeated reactions. The half-life of the free cells was 2 days, while that of the entrapped cells was estimated to be 63 days. Thus, immobilization greatly improved the operational stability of the hydrolytic enzyme in the yeast cells. Optical purity of the product was also constantly maintained as 100% after a long-term operation.

Figure 10 shows a diagram for a large scale production of *l*-menthol by the use of immobilized *R. minuta* var. *texensis* cells. Starting material, *dl*-menthol was converted into the succinate ester, then hydrolyzed stereospecifically. The yield was 86%. *d*-Menthyl succinate remained and succinic acid liberated were recycled as shown in this figure.

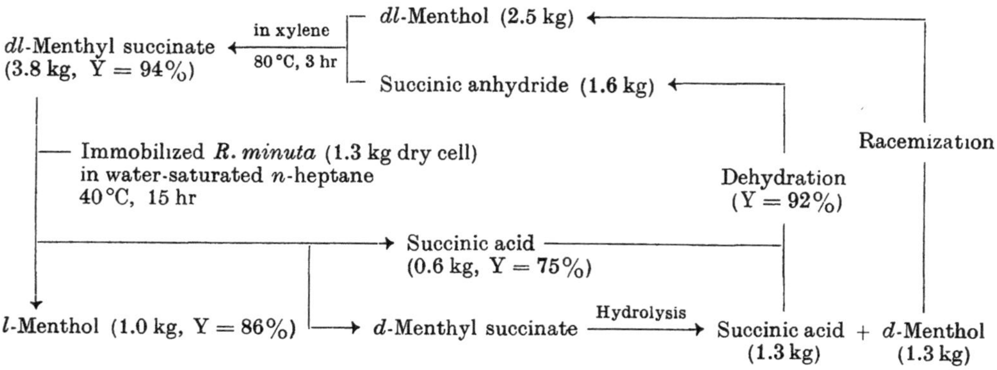

Fig. 10. Flow sheet of *l*-menthol production by gel-entrapped cells of *Rhodotorula minuta* var. *texensis*.

Conclusion

(1) The prepolymer methods to use photo-crosslinkable resin prepolymers or urethane prepolymers can easily make gel-entrapped biocatalysts of desired hydrophobicity, which can be applicable for the bioconversion of hydrophobic compounds in organic solvent systems.
(2) Bioconversion of steroids and a terpenoid in organic solvents could be achieved successfully by gel-entrapped microbial cells whose stability was improved by immobilization.
(3) Enzymes in microbial cells were active and stable in non-polar solvents. Therefore, the reaction should be carried out in less polar solvent systems so far as the solubility of substrates and products is enough.
(4) In less polar solvents, hydrophobicity of gels entrapping microbial cells affected seriously the transformation activity. Use of hydrophobic gels is preferable for bioconversion of lipophilic compounds, especially those with high hydrophobicity.
(5) The activity of gel-entrapped cells correlated closely to partition of substrates between gels and external solvents. Thus, selection of suitable gel hydrophobicity makes it possible to control conversion routes in the reactions which involve two or more reactants of different hydrophobicity.
(6) The results mentioned above are applicable for the bioconversion of various water-insoluble, lipophilic compounds.

Eingegangen: 23. 1. 81

References

[1] KLIBANOV, A. M., SAMOKHIN, G. P., MARTINEK, K. and BEREZIN, I. V.: Biotechnol. Bioeng. **19** (1977) 1351.
[2] BUCKLAND, B. C., DUNNILL, P. and LILLY, M. D.: Biotechnol. Bioeng. **17** (1975) 815.
[3] KLIBANOV, A. M.: Anal. Biochem. **93** (1979) 1.
[4] SONOMOTO, K., TANAKA, A., OMATA, T., YAMANE, T. and FUKUI, S.: Eur. J. Appl. Microbiol. Biotechnol. **6** (1979) 325.
[5] SONOMOTO, K., JIN, I.-N., TANAKA, A. and FUKUI, S.: Agric. Biol. Chem. **44** (1980) 1119.
[6] FUKUI, S., SONOMOTO, K., ITOH, N. and TANAKA, A.: Biochimie **62** (1980) 381.
[7] TANAKA, A., HAGI, N., YASUHARA, S. and FUKUI, S.: J. Ferment. Technol. **56** (1978) 511.

[8] Fukui, S., Tanaka, A., Iida, T. and Hasegawa, E.: FEBS Lett. **66** (1976) 179.
[9] Tanaka, A., Yasuhara, S., Osumi, M. and Fukui, S.: Eur. J. Biochem. **80** (1977) 193.
[10] Omata, T., Tanaka, A., Yamane, T. and Fukui, S.: Eur. J. Appl. Microbiol. Biotechnol. **6** (1979) 207.
[11] Fukushima, S., Nagai, T., Fujita, K., Tanaka, A. and Fukui, S.: Biotechnol. Bioeng. **20** (1978) 1465.
[12] Tanaka, A., Jin, I.-N., Kawamoto, S. and Fukui, S.: Eur. J. Appl. Microbiol. Biotechnol. **7** (1979) 351.
[13] Yamane, T., Nakatani, H., Sada, E., Omata, T., Tanaka, A. and Fukui, S.: Biotechnol. Bioeng. **21** (1979) 2133.
[14] Omata, T., Iida, T., Tanaka, A. and Fukui, S.: Eur. J. Appl. Microbiol. Biotechnol. **8** (1979) 143.
[15] Omata, T., Tanaka, A. and Fukui, S.: J. Ferment. Technol. **58** (1980) 339.
[16] Fukui, S., Omata, T., Yamane, T. and Tanaka, A.: Enzyme Engineering **5** (1980) 347.
[17] Fukui, S., Ahmed, S. A., Omata, T. and Tanaka, A.: Eur. J. Appl. Microbiol. Biotechnol. **10** (1980) 289.
[18] Omata, T., Iwamoto, N., Kimura, T., Tanaka, A. and Fukui, S.: Eur. J. Appl. Microbiol. Biotechnol. **11** (1981) 199.

Acta Biotechnologica 1 (1981) 4, 351–364

Ethanol Production from Non-Grain Feedstocks

St. L. MICHAELS[1] and H. W. BLANCH[2]

1 TERA Corporation
2150 Shattuck Ave., Suite 1200
Berkeley, CA 94704, USA

2 Department of Chemical Engineering
University of California
Berkeley, CA 94720, USA

Summary

A general view of the possibilities of producing ethanol from sugar, starch and cellulose feedstocks is given.
For the 3 variants net energy analysis of ethanol production and evaluation of costs are presented. With the exception of the case using molasses as feedstock the net energy balances are positive.
The greatest possible net energy yield can be expected with sugar cane followed by sugar beets, wood and paper waste. Based on feedstock availability, net energy utilization and production costs, the most promising processes for producing ethanol from non-grain feedstocks over the next 20 years will be those processes using fermentable sugars available from nongrain starchy materials, cellulosics and whey.
The feedstock prices for cellulosics are low and if the developments in cellulose hydrolysis will lead to improve the ethanol yields from cellulose fermentation to nearer 90 percent of the theoretical value, cellulosic materials can become a good feedstock for ethanol production.

Zusammenfassung

Ein allgemeiner Überblick über die Herstellungsmöglichkeiten von Äthanol aus Zucker, Stärke und Zellulose wird gegeben. Für diese drei Varianten werden die Analyse der Nettoenergie der Äthanolproduktion und die Kostenentwicklung aufgezeigt. Mit Ausnahme der Melasseverwertung sind die Nettoenergiebilanzen positiv.
Die höchstmögliche Nettoenergieausbeute kann bei der Verwendung von Zuckerrohr, gefolgt von Zuckerrüben, Holz und Abfallpapier erwartet werden. Unter Berücksichtigung der Verfügbarkeit der Rohstoffe, der Nettoenergienutzung und der Produktionskosten können als die aussichtsreichsten Prozesse zur Herstellung von Alkohol aus anderen Rohstoffen als Getreide in den nächsten 20 Jahren solche angesehen werden, die fermentierbare Zucker aus stärkehaltigen Produkten, außer Getreide, aus Zellulose und aus Molke verwenden.
Die Rohstoffpreise für zellulosehaltige Materialien sind niedrig; falls die Entwicklung der Zellulosehydrolyse dazu führt, daß die Äthanolausbeuten bei 90% des theoretischen Wertes liegen, können zellulosehaltige Materialien wertvolle Rohstoffe für die Äthanolherstellung werden.

Introduction

Ethanol is the primary liquid fuel which can be manufactured from renewable resources through available commercial technology. In the United States, ethanol is currently manufactured by the fermentation of glucose obtained by saccharifying corn starch. The anhydrous ethanol thus produced is blended with gasoline for sale as unleaded automotive fuel. Ethanol can also be blended with diesel fuel or can be burned to fuel gas turbines or steam boilers. Although ethanol produced from biomass is currently too expensive to penetrate the markets for gas turbine or boiler fuels [1], the high prices and uncertain supplies of gasoline and diesel fuels have opened a widespread market for ethanol as a premium transportation fuel. In the United States, the potential demand for ethanol to be blended into a five percent "gasohol" nationwide is 5.81 billion gallons per year [2]. If the proposed 1990 U.S. goal of ten percent "gasohol" nationwide is to be attained, ethanol demand will approach 11.6 billion gallons per year [3].

Table 1. Availability and Cost of Non-Grain Feedstocks for Ethanol Production in the U.S.A.

Feedstock	Available Mass (tonnes/year)	Anhydrous Ethanol Equivalent (m^3/year)	Cost (1979 $/tonne)
Sugar Cane	25,900 [4]	1,710	21.10 [5]
Molasses	imported	0.316 m^3/tonne	93.70 [7]
Sugar Beets	23,300,000 [4]	1,912,530	28.50 [5]
Sweet Sorghum	242,000 [4]	21,600	11.55 [8]
Citrus Wastes	1,727,000 [8*]	774,200	85.00 [8*]
Cheese Whey	818,200 [8*]	587,100	41.65 [8*]
Culled and Waste Potatoes	1,835,000 [6]	214,700	22.00
Cassava	none	0.174 – 0.265 m^3/tonne [9]	36.63 [10]
Wood	375,000,000 [11*]	108,000,000	35.50 [8*]
Waste Paper	12,700,000 [8*]	4,170,000	26.40 [12*]
Agricultural Residues	373,854,000 [13*]	95,000,000	40.48 [8*]

* dry weight

Ethanol can be manufactured from monosaccharides such as glucose, fructose, galactose, and xylose, disaccharides such as sucrose and lactose, and polysaccharides such as starch, hemicellulose, and cellulose. Potential feedstocks and their current availability and cost in the U.S. are listed in Table 1. Because corn and other grains are important livestock feeds, implementation of substantial grain-based ethanol production risks straining food supplies and increasing grain costs — and thus ethanol cost — in world markets [14]. For this reason, cheaper, more abundant non-grain feedstocks will become increasingly important raw materials for alcohol production.

General Process Description

Three common classes of chemical processes comprise ethanol production technology: saccharification, fermentation, and product recovery. Figure 1 shows how these processes interact and lists some of the specific unit operations in each class. In saccharification, the raw material is processed to extract the fermentable compounds, which are

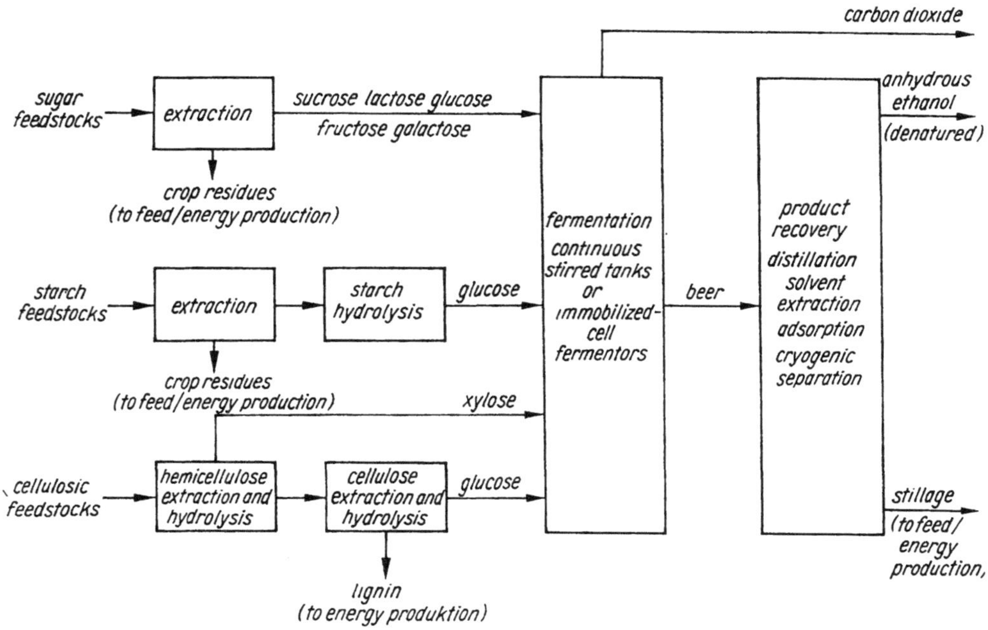

Fig. 1. Unit operations of ethanol manufacture

then chemically modified as necessary to yield either monosaccharides or disaccharides which can be directly fermented by the chosen fermentation microorganisms.

In the fermentation, mesophilic yeast (*Saccharomyces cerevisiae* or *Kluyveromyces fragilis*) or thermophilic bacteria (*Clostridium thermocellum* and *Clostridium thermosaccharolyticum*) are grown anaerobically in a sterilized aqueous solution of the sugars,

Table 2. Energy Requirements for Ethanol Recovery Processes

Recovery Process	Energy Requirement (GJ/m³)	Developmental Status
Conventional		
Atmospheric Distillation (to 95% ethanol)	6.996	
Extractive Distillation (to anhydrous ethanol)	2.090	
TOTAL	9.086 [7]	Commercial
Advanced		
Multi-Pressure Distillation	5.957 [15]	Commercial
Vacuum Distillation	3.570 [7]	Commercial
Cryogenic Separation	2.500*	Hypothetical
Distillation/Adsorption	2.550 [16]	Developmental: Demonstrated
– Calcium Oxide Adsorbent		
– Cellulose Adsorbent	1.997 [16]	Developmental: Demonstrated
Solvent Extraction with Ether or Gasoline	1.004**	Hypothetical

* Estimated from latent heats of melting of ethanol and water
** Estimated from distillation of diethyl ether/ethanol binary

which are metabolized under ideal conditions to equal moles of ethanol and carbon dioxide. By-products include cells mass and unfermented sugars in a dilute solution called stillage, and other low molecular weight aldehydes and alcohols called fusel oils.
Modern fermentors are available for batch or continuous operation. High-yield modern fermentor designs incorporate continuous recycle to the fermentor of the fermentation organisms separated from the ethanol solution. Cell recycle maintains a high cell density and permits a short fermentation residence time. Immobilized-cell reactors are under

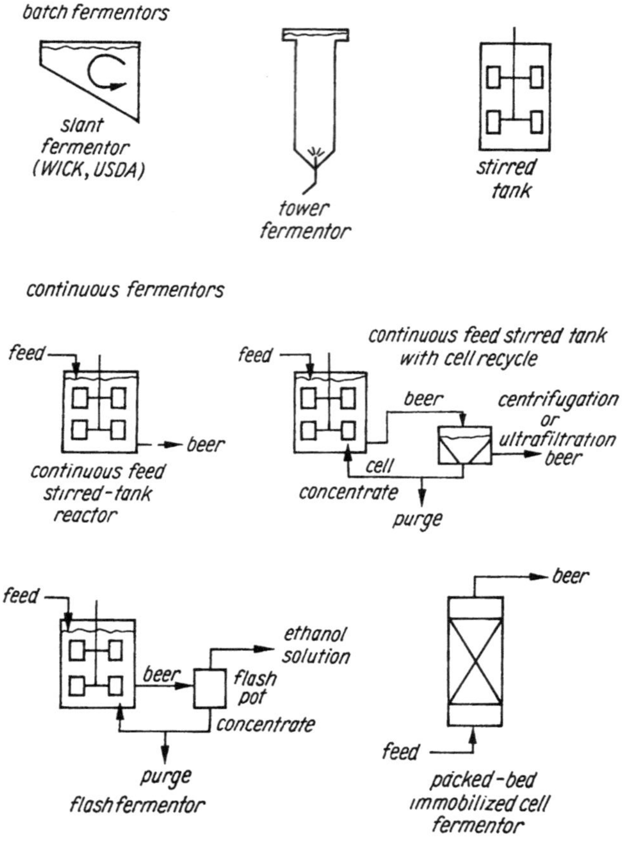

Fig. 2. Fermentor configurations

development in which cells are grown immobilized on a porous packed support which permits feed and product to diffuse to and from the cells; these reactors are operated in a plug-flow mode. Figure 2 illustrates some modern fermentor configurations.
Ethanol is recovered conventionally by distillation to the 95 volume percent aqueous azeotrope, followed by an extractive distillation with benzene, hexane, cyclohexane, or gasoline [7]. Multiple-pressure stills are also available which require no extracting solvent to yield anhydrous ethanol [15]. Under development are processes to adsorb water onto calcium oxide or cellulose adsorbents [16], ethanol extraction with ethers or gasoline, and cryogenic separations. The goal of new developments in ethanol recovery is the reduction of the high energy requirements for distillation. Energy requirements estimated for some of these ethanol recovery processes are compared in Table 2.

Ethanol Manufacture from Sugars and Sugar Crops

Monosaccharides and disaccharides obtained from such diverse sources as sugar cane, sweet sorghum, sugar beets, molasses and cheese whey can be fermented to ethanol. These common fermentable sugars include glucose, fructose, galactose, sucrose, and lactose. If the raw material is not already in the form of a concentrated aqueous sugar solution like molasses or concentrated whey permeate, the material must be crushed and washed to remove the sugars and the ensuing dilute sugar solution must be concentrated to about 15 weight percent. Equipment such as continuous countercurrent extractors presses and evaporators are used to extract and concentrate these sugars commercially. The sugar solutions must sometimes be demineralized or deproteinized to optimize the fermentation; these steps are required for beet sugar and whey. The sugar solutions are sterilized, mixed with vitamins, minerals and nitrogen and phosphate sources, and fed to the fermentors where the sugars are converted to equimolar amounts of ethanol and carbon dioxide by the yeasts *S. cerevisiae* (glucose, fructose and sucrose fermentations) or *K. fragilis* (lactose, glucose and galactose fermentations). Yields of ethanol typically are 45 – 50 percent of sugar weight. The yeast cells are removed from the fermentation beer by centrifugation or ultrafiltration and are recycled to the fermentor. The remaining clarified beer is distilled or otherwise processed to dehydrate the ethanol to a 95 volume percent aqueous azeotrope or an anhydrous fuel-grade stream often denatured with gasoline or the traces of fusel oils co-produced during the fermentation.
By-products from these processes include crop wastes such as sugar cane bagasse, beet pulp, or sorghum pith, aqueous stillage containing the plant's net yeast cell production mixed with the unfermented sugar solution from the still bottoms, carbon dioxide, and fusel oils. The crop wastes can be burned to generate power and process heat for the plant or can be fermented themselves in a cellulose fermentation process. Beet pulp makes excellent livestock feed as does the mixture of yeast and still bottoms, which must be dried. If no local feed market exists, these by-products can also be combusted in a dilute solution by wet air oxidation to generate process heat for the plant [17]. Fusel oils are added to the anhydrous ethanol as denaturing agents. Carbon dioxide can be vented to the atmosphere or can be compressed or frozen for sale to a local customer. Values of these by products are listed in Table 3.

Table 3. Value of By-Products of Ethanol produced from Sugar Feedstocks

By-Product	Use	Value	By-Product Credit ($/m³ ethanol)
Bagasse	Fuel	$ 27.30/1000 kg [4]	used internally
	Ethanol Production	$ 24.89/1000 kg*	29.75
Beet Pulp	Livestock Feed	$ 173/1000 kg [4]	12.05
Sorghum Pith	Fuel	$ 23.20/1000 kg [4]	used internally
Whey Protein	Food Additive	$ 2,640/1000 kg**	572.00
Stillage***	Livestock Feed	$ 992/1000 kg	59.44
Carbon Dioxide	Food Uses	up to $ 440/1000 kg [18]	up to 382.20
Fusil Oils	Ethanol Denaturant	none	used internally

* Based on ethanol at $ 422/m³ minus production costs of $ 330/m³
** Based on $ 1.32/kg value for whey protein
*** Includes yeast mixed with still bottoms and dried [7]

Figure 3 demonstrates the important energy flows in and out of a typical process to ferment sugar crops to ethanol. Table 4 lists the net energy analyses for processes to ferment sugar cane, cane molasses, sweet sorghum, sugar beets, and whey. The least energy-intensive process is whey fermentation which consumes a waste product carrying a low energy burden from its production. Molasses fermentation, on the other extreme, incurs a net energy loss, while other sugar crops show a net yield because the crop wastes can be used as feed or fuel.

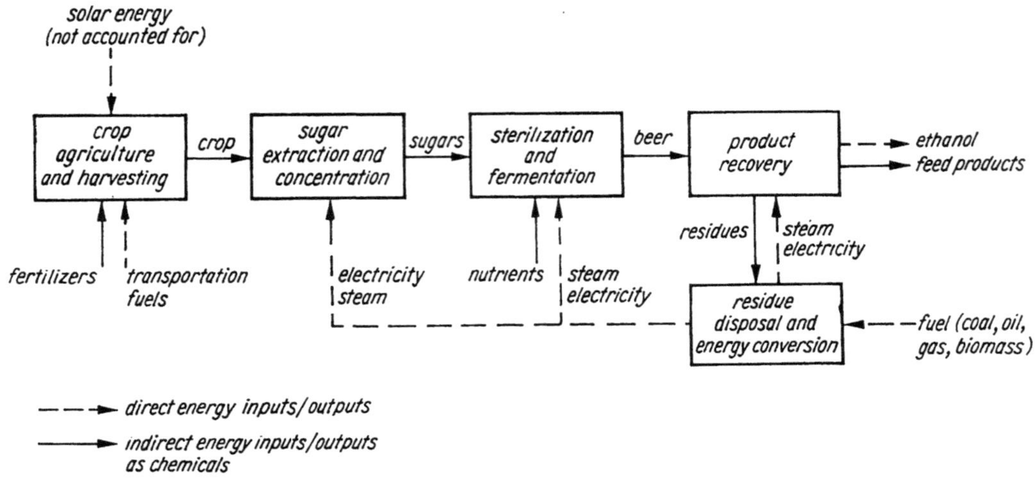

Fig. 3. Energy flows in sugar fermentations

Table 5 lists the major components of production costs for fermenting sugar feedstocks to anhydrous ethanol. Costs are highly dependent on raw material cost and process energy requirements, as the overall plant equipment does not vary widely for different materials in this class of feedstocks. The least expensive raw material is whey because of its characterization as a low value waste stream; even with extra value credited to the

Table 4. Net Energy Analysis of Ethanol Production from Sugar Feedstocks (MJ/m^3 Anhydrous Ethanol)

Feedstock	DEBIT: Energy to Produce Feedstock	DEBIT: Ethanol Process Energy Requirement	CREDIT: Ethanol Product	CREDIT: By-Product Energy*	NET Energy Yield
Sugar Cane [19]	4,680	12,230	21,210	19,800	24,200
Molasses [7]	14,530	9,620 [7]	21,210	–	–2,940
Sweet Sorghum [19]	6,260	12,880	21,210	14,920	16,990
Sugar Beets [4]	10,220	12,500	21,210	29,410	23,870
Whey	10**	9,620	21,210	–	11,580

* Energy of combustion of crop wastes

** Estimate of energy to concentrate whey permeate to 15 weight percent solids by evaporation and transport 65 km by truck. No energy debit is made for whey permeate production as agricultural energy input is allotted to milk products not milk wastes.

whey producer for lactose, whey fermentation produces the lowest cost ethanol among the sugar feedstocks. No reliable cost estimates are available for fermentation of citrus wastes.

Table 5. Cost of Ethanol Production from Sugar Feedstocks (1979 $)
Basis 95 000 m^3/years Anhydrous Ethanol

Feedstock	Feedstock Cost ($/$m^3$)	Process Operating and Maintenance Costs ($/$m^3$)	Capital Amortization Insurance and Taxes ($/$m^3$)	Extractive Distillation Costs ($/$m^3$)	Total ($/$m^3$)
Sugar Cane	267.90[a]	85.20[b]	35.80[c]	9.50[d]	398.40
Molasses	331.80[e]	65.30	17.10	9.50	423.70
Sweet Sorghum[f]	131.80	85.20	35.80	9.50	262.30
Sugar Beets[f]	347.20	85.20	35.80	9.50	477.70
Whey[g]	128.30[h]	65.30	17.10	9.50	220.20

[a] Cane cost $ 21.10/1000 kg.
[b] O & M costs for milling $ 19.87/m^3. [4] and for fermentation $ 65.32/m^3 [7].
[c] Capital costs $ 18.68/m^3 for milling based on 13.7% annualization factor applied to $ 9.9 million plant producing 411 × 10^3 kg/day fermentable sugars Capital cost derived from $ 30.5 million plant producing 2,665,00 kg/day of fermentable sugars by 0.6 scale factor [4]. Capital costs $ 17.12/m^3 for fermentation based on 13.7% annualization factor applied to $ 14.4 million plant producing 95,000 m^3/yr.
[d] Based on dehydration column capital cost of $ 400,000 and steam load of 2,090 MJ/m^3 at $ 4.27/GJ energy cost.
[e] Reference 7, p. 25.
[f] Sorghum and beet milling and fermentation costs are comparable with cane milling and fermentation costs.
[g] Whey fermentation costs are nearly identical to molasses fermentation costs.
[h] Costs of whey permeate concentration to 15% lactose and transportation 40 miles plus 4.4 cents/kg lactose value to complete with lactose recovery.

Ethanol Manufacture from Non-Grain Starches

Starches, glucose polymers characterized by the α-1,4 glycosidic linkage between glucose residues, are available from such non-grain feedstocks as potatoes and cassava (also known as manioc or tapioca). These starchy tubers are sliced and slurried with water, then mixed with amylase and amyloglucosidase, fungal enzymes which respectively hydrolyze the starch into oligomers and the oligomeric polysaccharides into glucose. These enzyme additives cost $ 7.14/m^3 and $ 10.74/m^3 ethanol produced, respectively. The glucose solution is then sterilized, mixed with nutrients, and fermented by *S. cerevisiae* as described previously. Ethanol and carbon dioxide are again produced in equimolar quantities for ethanol yields of 45—50 percent of starch weight.

By-products from starch fermentation processes include very small amounts of feedstock residues such as cassava fibers, stillage, carbon dioxide, and fusel oils. These by-products can be processed as previously described to yield energy for the plant, livestock feed supplements, liquid or solid carbon dioxide for sale, and denaturants for fuel-grade ethanol. Values of these by-products are listed in Table 6.

Figure 4 illustrates the important energy flows in and out of a typical starch fermentation process. Table 7 lists the net energy analyses for processes to ferment potatoes or potato wastes, cassava, and manioc. Starch-based ethanol processes produce less net energy than sugar or cellulose-based processes because the starch feedstocks contain little fiber which can be burned or sold to produce a by-product energy flow.

Table 6. Value of By-Products of Ethanol produced from Starch Feedstocks

By-Produkt	Use	Value	By-Product Credit ($/m³ ethanol)
Cassava Skin & Fibers	Fuel	$ 16.76/1000 kg*	used internally
Stillage	Livestock Feed	$ 992/1000 kg	59.44
Carbon Dioxide	Food Uses	up to $ 440/1000 kg [18]	up to 382.20.
Fuel Oils	Ethanol Denaturant	none	used internally

* Based on $ 0.94/GJ energy value when combusted. Yield 426.4 kg/m³ ethanol [10]

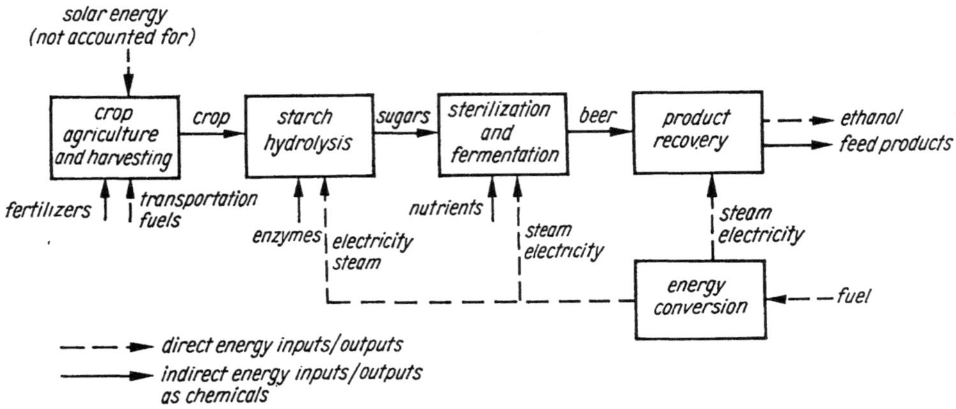

Fig. 4. Energy flows in starch fermentations

Table 8 lists the major components of production costs for fermenting starch feedstocks to fuel-grade anhydrous ethanol. Costs are again highly dependent on raw material cost and process energy requirements. Starch fermentations are seen to be similar in production costs to sugar cane fermentation, and more expensive than producing ethanol from sweet sorghum or whey. Again, the use of wastes or high-yield crops with low food values tends to reduce feedstock cost and net energy requirements, and thus improve the economics and energy consumption associated with ethanol manufacture.

Table 7. Net Energy Analysis of Ethanol Production from Starch Feedstocks (MJ/m³ Anhydrous Ethanol)

Feedstock	DEBIT: Energy to Produce Feedstock	DEBIT: Ethanol Process Energy Requirement**	CREDIT: Ethanol Product	CREDIT: By-Product Energy	NET Energy Yield
Potatoes	—*	14,210	21,210	—	7,000
Cassava [19]	4,110	14,200	21,210	—	2,900

* Culled potatoes used at site of potato processor; energy to grow potatoes is debited to the food products not the potato wastes and culls.

** Reference 10, p. 17. Extra energy for potato processing is required for removol of the larger amount of moisture contained in potatoes.

Table 8. Cost of Ethanol Production from Starch Feedstocks (1979)
Basis 49,500 m³/year Anhydrous Ethanol

Feedstock	Feedstock Cost ($/m³)	Process Operating and Maintenance Costs ($/m³)	Capital Amortization, Insurance and Taxes ($/m³)	Extractive Distillation Costs ($/m³)*	Total ($/m³)
Cassava [10]	228.00	91.00	47.00	9.50	375.50
Potatoes**	83.00***	107.00	47.00	9.50	246.50

* Based on dehydration column capital cost of $ 250,000 and steam load of 2090 MJ/m³ at $ 4.27/GJ energy cost.
** Plant configuration and size for potato fermentation is very similar to that for cassava fermentation but O & M costs are higher because of higher water content of potatoes which must be removed.
*** Based on culled potatoes at $ 22.00/1000 kg.

Ethanol Manufacture from Cellulose

Cellulose, a glucose polymer characterized by the β-1, 4, glycosidic linkage between glucose residues, and hemicellulose, a polymer of the five-carbon sugar xylose, are the major polysaccharide constituents of wood, paper, and agricultural residues. The polysaccharides are bound into a chemical matrix by lignin, a phenylpropane network polymer which protects the cellulose from enzymatic or chemical attack under ambient conditions. If the polysaccharides are extracted from the lignin, usually by chemical and physical disruption of the lignocellulose, they can be depolymerized to glucose and xylose in a hydrolysis stage. The resulting monosaccharides can then be fermented to ethanol in a conventional fermentation.

Extraction and hydrolysis can occur in several discrete unit operations or in one consolidated operation depending upon the chosen hydrolysis method. Figure 5 shows these processing operations and their interaction for several methods of extraction and hydrolysis. Hemicellulose is normally extracted, hydrolyzed and solubilized under dilute acid and low temperature conditions with sulfuric, [20] nitric [21], or hydrochloric acid as the catalyst. Cellulose is much harder to extract and solubilize, and is normally processed

Table 9. Theoretical Yield of Ethanol from Cellulosic Feedstocks (Assuming Conversion of Both Xylose and Glucose)

Feedstock	Yield per Oven-Dry Tonne	
	(kg)	(m³)
Oak	324	0.41
Pine	302	0.38
Fir	330	0.42
Bagasse	303	0.38
Oat Hulls	312	0.40
Rice Straw	279	0.35
Wheat Straw	288	0.36
Corn Cobs	313	0.40
Corn Stover	293	0.37
Waste Paper	389	0.49

with hot concentrated acids and steam or enzymes produced by a range of fungi and bacteria (*Trichoderma viride, Clostridium thermocellum*) [20]. Lignin has no current fermentation value, although lignin-consuming organisms are under study and development. The lignin remains chemically unmodified by hydrolysis, and can be recovered in a solid or slurry form for further processing or disposal.

Most advanced cellulose fermentation processes extract and hydrolyze hemicellulose and cellulose in consecutive steps to yield separate glucose and xylose solutions [20]. The

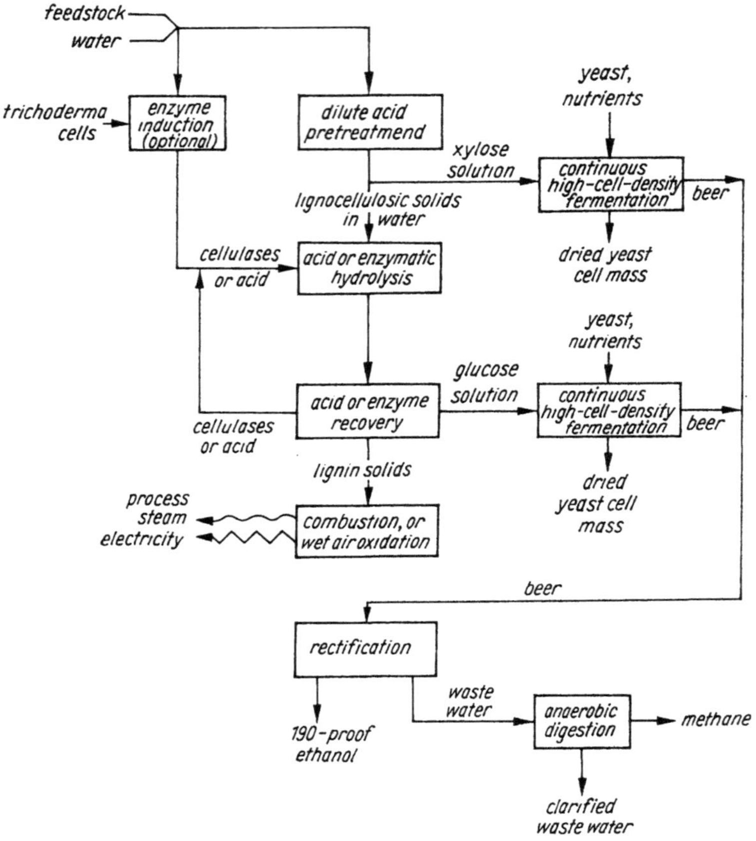

Fig. 5. Process operations in cellulose fermentations

glucose solution is then fermented to ethanol by *S. cerevisiae* or *C. thermosaccharolyticum* while the xylose solution can be fermented to ethanol by *C. thermosaccharolyticum* or *Bacillus mascerans*. Maximum theoretical ethanol yields from cellulosic feedstocks are listed in Table 9; advanced fermentation processes have attained 60–70 percent of these theoretical yields. Advanced fermentation processes may also use selected mixed bacterial cultures in which one organism can hydrolyze hemicellulose and cellulose, while another ferments glucose and xylose simultaneously to produce ethanol. Examples of these simultaneous saccharification/fermentation processes include the Gulf process [3], General Electric/University of Pennsylvania process [2], and MIT process [25]. The latter two processes employ mutant strains of the thermophilic bacteria *C. thermocellum* and *C. thermosaccharolyticum* which have been selected for high ethanol productivity and low acetic acid/acetone productivity. Again, the fermentations may be carried out in batch or continuous stirred tank fermentors with or without cell recycle, or in immobilized-cell fermentors. Product recovery options are the same as for other ethanol fermentation processes mentioned previously.

By-products from cellulose fermentation processes include lignin, stillage, carbon dioxide, fusel oils, and sugar degradation by-products resulting from some methods of cellulose hydrolysis. Lignin is the most plentiful by-product. Although lignin can be reacted to produce an economically viable lignin-formaldehyde binding resin most current plant designs intend to combust the lignin by wet air oxidation [21] or dry combustion [26] to generate process heat for the ethanol plant. Sugar by-products and stillage can again be

Table 10. Value of By-Products of Ethanol produced from Cellulosic Feedstocks

By-Product	Use	Value	By-Product Credit ($/m³ ethanol)
Lignin	Fuel	$ 25.30/1000 kg*	used internally
	Formaldehyde Binder	$ 440.00/1000 kg	275
Stillage	Feed/Fertilizer	$ 992/1000 kg	59.44
Carbon Dioxide	Food Uses	up to $ 440/1000 kg [18]	382.20
Fusel Oils	Ethanol Denaturant	none	used internally

* Based on $ 0.94/GJ energy value

sold as a livestock feed supplement, can be combusted by wet air oxidation, or can be angerobically digested to methane, a potential boiler fuel for the ethanol plant [21, 26] Carbon dioxide can be vented or sold, and fusel oils can serve as ethanol denaturants. Values of these by products are listed in Table 10.

Figure 6 illustrate the important energy flows in and out of a typical cellulose fermentation process. Table 11 lists the net energy analyses for processes to ferment various cellulosic raw materials.

Table 12 lists the major components of production costs for fermenting cellulose feedstocks to fuel-grade anhydrous ethanol. Costs are dependent on costs of raw materials and other chemical feedstocks required, as well as process energy requirements. Compared to fermentation of sugar or starch feedstocks, cellulose fermentations require more energy,

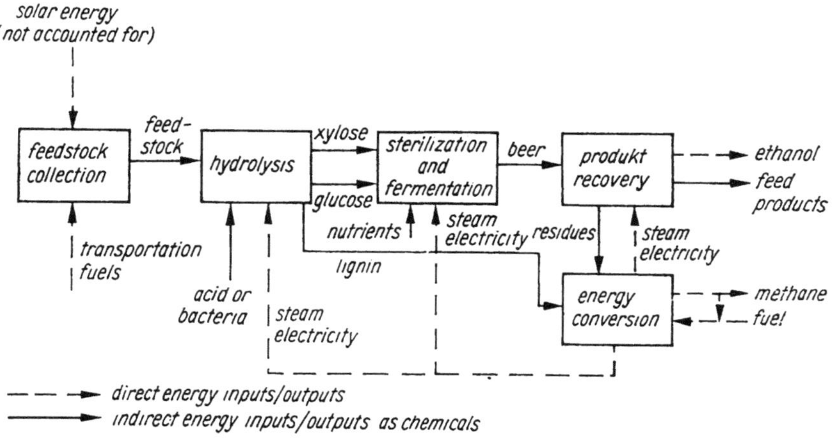

Fig. 6. Energy flows in cellulose fermentations

Table 11. Net Energy Analysis of Ethanol Production from Cellulosic Feedstocks (MJ/m^2 Anhydrous Ethanol)

Feedstock	DEBIT: Energy to Produce Feedstock	DEBIT: Ethanol Process Energy Requirement[a]	CREDIT: Ethanol Product	CREDIT: By-Product Energy	NET Energy Yield
Wood (Fir)	10[b]	14,650	21,210	17,160[c]	23,710
Waste Paper	370[d]	14,650	21,210	17,070[e]	23,260
Agricultural Residues	10	14,650	21,210	11,120[f]	17,670

[a] Reference 7 plus 5,030 MJ/m^3 for cellulose and hemicellulose hydrolysis.
[b] Additional energy to collect and transport culled wood, wood chips, or agricultural residues during normal harvesting operations.
[c] Based on composition of 27% lignin and heat of combustion of 11,500 Btu/lb lignin [27].
[d] Energy to separate municipal solid waste into fermentable/combustible and non-combustible components is 50 kWh/ton [27].
[e] Based on composition of 21% lignin [28].
[f] Based on composition of 15% lignin [21].

but this requirement is offset by the much lower agricultural energy requirements to produce wood or residues, and the energy credit obtained from combustion or other conversion of the lignin by-product. If the recent process developments in cellulose hydrolysis and ethanol yield improvement can be brought to commercial fruition, the processes based on cellulose will be able to compete with ethanol processes based on the more expensive sugar feedstocks. Because agricultural wastes are abundant and will not be subject to the high inflation rates which can affect sugar and grain crops under conditions of tight supply, production of ethanol from non-food feedstocks such as wood and agricultural wastes will minimize the impact of large scale ethanol production on world food and energy supplies while permitting the maximum production of ethanol at minimum price as a liquid fuel or chemical feedstock.

Based on feedstock availability, net energy utilization and production costs, therefore, the most promising processes for manufacturing fuel-grade ethanol from non-grain feedstocks over the next 20 years will be those processes fermenting the sugars available from cellulosic materials, non-grain starchy materials and whey. The high productivity ethanol processes currently under development will greatly improve the energy efficiency of ethanol production, but will probably decrease production costs only slightly because of the large feedstock price component in ethanol production costs. These developmental processes will be able to realize only slight improvements in the already-high yields from

Table 12. Cost of Ethanol Production from Cellulosic Feedstocks (1979 $) Basis 122,600 m^3 years Anhydrous Ethanol

Feedstock	Feedstock Cost ($/m³)	Process Operating and Maintenance Costs ($/m³)	Capital Amortization, Insurance and Taxes* ($/m³)	Extractive Distillation Costs ($/m³)	Total ($/m³)
Wood	52.40	277.00	48.10	9.50	387.00
Waste Paper	53.90	277.00	43.20	9.50	383.60
Agricultural Residues	103.10	277.10	58.10	9.50	447.80

* Based on $ 43 million cost for 122,600 m^3/yr from dry fir [26]. Costs adjusted for different yields.

sugars and starches, but will substantially improve the ethanol yields from cellulose fermentations from 60—70 percent of theoretical to nearer 90 percent, decreasing the cost of ethanol production from wood and agricultural residues. Because of these developments, as the cost of energy increases, and the availability of grain for fuel production decreases, ethanol produced from non-grain feedstocks will compete successfully with ethanol produced from grain and may succeed in replacing grain ethanol as the major fuel alcohol in the U.S.

Eingegangen: 20. 1. 81

References

[1] BENEMANN, J. R., Biofuels: A Survey. Palo ALTO, California; Electric Power Research Institute, ER-746-SR, June, 1978.
[2] PARK, W., PRICE, G., and SALO, D., Biomass-Based Alcohol Fuels: The Near-Term Potential for Use with Gasoline, Washington, D.C.: NTIS HCP/14101-03, August, 1978, p. 25.
[3] U.S. Senate Bill 1308 (pending).
[4] NATHAN, R. A., ed., Fuels from Sugar Crops, Washington, D.C.: NTIS TID-22781, 1978.
[5] "Agricultural Prices", Washington, D.C.: U.S. Department of Agriculture, Crop Reporting Board, July, 1979.
[6] "Agricultural Statistics", Washington, D.C.: U.S. Department of Agriculture, 1979, p. 181.
[7] MAIORELLA, B., BLANCH, H. W., and WILKE, C. R., Rapid Ethanol Production via Fermentation, Berkeley, California: Lawrence Berkeley Laboratory, LBL-10219, November, 1979.
[8] The Report of the Alcohol Fuels Policy Review, Washington, D.C.: NTIS DOE/PE-0012, June, 1979.
[9] MARZOLA, D. L., and BARTHOLOMEW, D. P., "Photosynthetic Pathway and Biomass Energy Production", Science, vol. 205, no. 4406, (1979), pp. 555—559.
[10] YANG, V., and TRINDADE, S. C., "Brazil's Gasohol Program", Chemical Engineering Progress, vol. 75, no. 4, (1979), pp. 11—19.
[11] FALKEHAG, I., in Progress in Biomass Conversion, Saakamen, G., and D. Tillman, eds., vol. 1, New York: Academic Press, 1979, p. 1.
[12] HUMPHREY, A. E., and NOLAN, E. J., "Summary Report on the Biological Production of Liquid Fuels from Biomass", Third Annual Biomass Energy Systems Conference Proceedings, Washington, D.C.: NTIS SERI/TP-33-285, October, 1979, pp. 581—587.
[13] The Potential of Producing Energy from Agriculture, Ch. 2, "Agricultural Residues", Washington, D.C.: Office of Technology Assessment, U.S. Congress, May, 1979, pp. 5—34.
[14] Gasohol: A Technical Memorandum, Washington, D.C.: Office of Technology Assessment, U.S. Congress, September, 1979.
[15] Raphael KATZEN Associates, Grain Motor Fuel Alcohol Technical and Economic Assessment Study, Washington, D. C.: NTIS HCP/J6639-01, June, 1979.
[16] LADISCH, M. R., and DYCK, K. "Dehydration of Ethanol: New Approach Gives Positive Energy Balance", Science, vol. 205, no. 4409, (1979), pp. 898—890.
[17] WILHELMI, A. R., and KNOPP, P. V. "Wet Air Oxidation — An Alternative to Incineration", Chemical Engineering Progress, vol. 75, no. 8, (August, 1979), pp. 46—52.
[18] Dry ice price quotation from Union Ice Company, Oakland, California, May 15, 1980.
[19] DA SILVA, J. G., SERRA, G. E., MOREIRA, J. R., CONCALVES, J. C., and GOLDEMBERG, J., "Energy Balance for Ethanol Production from Crops", Science, vol. 201, no. 4359, (1978), pp. 903—6.
[20] WILKE, C. R., BLANCH, H. W., SCIAMANNA, A. F., ROSENBERG, S. L., TANGNU, S. K., and FREITAS, R. P., "Process Development Studies on the Bioconversion of Cellulose and Production of Ethanol", Third Annual Biomass Energy Systems Conference Proceedings, Washington, D.C.: NTIS SERI/TP-33-285, October, 1979, pp. 79—84.
[21] BRINK, D. L., and MERRIMAN, M. M., "Potential for Producing Alcohol from Organic Residues", Richmond, California: University of California Forest Products Laboratory, paper presented in Fresno, California, February 19, 1980.

[22] CONVERSE, A. O., and GRETHLEIN, H. E., "Acid Hydrolysis of Cellulosic Biomass", Third Annual Biomass Energy Systems Conference Proceedings, Washington, D. C.: NTIS SERI/ TP-33-285, October, 1979, pp. 581–587.

[23] EMERT, G. H., and KATZEN, R., "Chemicals from Biomass by Improved Enzyme Technology", Symposium Proceedings, Biomass as a Non-Fossil Energy Source, Washington, D.C.: American Chemical Society, April, 1979.

[24] PYE, E. K., and HUMPHREY, A. E., "Production of Liquid Fuels from Cellulosic Biomass", Third Annual Biomass Energy Systems Conference Proceedings, Washington, D.C.: NTIS SERI/TP-33-825, October, 1979, pp. 69–75.

[25] WANG, D. I. C., Biocic, I., Fang, H-Y, and S-D Wang, "Direct Microbiological Conversion of Cellulosic Biomass to Ethanol:, ibid., pp. 61–67.

[26] WILKE, C. R., YANG, R. D., SCIAMANNA, A. S., and FREITAS, R. P., "Raw Materials Evaluation and Process Development Studies for Conversion of Biomass to Sugars and Ethanol", Second Annual Symposium on Fuels from Biomass, Troy, N. Y.: Rensselaer Polytechnic Institute, June 20, 1978, pp. 421–459.

[28] WILKE, C. R., Cellulose, Food, and Energy, Berkeley, California: Lawrence Berkeley Laboratory, LBL-5275, November, 1977.

[29] WEISZ, P. B., and MARSHALL, J. F., "High-Grade Fuels from Biomass Farming: Potentials and Constraints," Science, vol. 206, no 4414, (1979), pp. 24–29.

[30] Teknekron, Inc., Pollutant Releases, Resource Requirements, Costs, and Efficiencies of Selected New Energy Technologies, Berkeley, California: Teknekron, Inc., December, 1975.

Acta Biotechnologica 1 (1981) 4, 365–370

Bioconversion of Cellulose Wastes to Protein and Sugar

V. R. SRINIVASAN and T. F. MILLER

Department of Microbiology
Louisiana State University
Baton Rouge, LA 70803

Paper given at the 2nd Symposium of Socialist Countries on Biotechnology, Leipzig 2.–5. 12. 1980

Summary

A schematic representation of the variety of products which can be obtained by microbial conversion of cellulose is presented.
Alkaline pre-treatment has been used after milling in all the experiments. Solka-floc or sugarcane bagasse was used as sources of cellulose. A cellulolytic strain of *Aspergillus terreus* (ATCC 30514) was cultivated in batch-, fed batch and continuous culture up to 7 liter stirred tank fermenter. The general growth characteristics were determined by growing on glucose. Results of experiments on the growth of *A. terreus* for production of biomass on Solka-floc or Sugarcane bagasse are given, also the ability of crude cellulases to produce sugar syrups by enzymatic hydrolysis of cellulose has been evaluated.

Zusammenfassung

Einleitend wird ein Schema über die Vielfalt der Produkte, die durch mikrobielle Umwandlung von Cellulose erhalten werden können, vorgestellt.
Bei allen experimentellen Untersuchungen wurde eine Mahlung und alkalische Vorbehandlung der cellulosehaltigen Substrate (Solka-floc und Zuckerrohr-Bagasse) vorgenommen. Der cellulolytisch aktive Stamm *Aspergillus terreus* (ATCC 30514) wurde diskontinuierlich, semikontinuierlich und kontinuierlich in Volumina bis zu 7 Liter (Laborrührfermentor) kultiviert. Die allgemeine Wachstumscharakteristik wurde bei Glucose-limitiertem Wachstum ermittelt.
Experimentelle Ergebnisse des Wachstums von *A. terreus* zur Biomasseproduktion auf den obengenannten Substraten als auch des Einsatzes der Rohenzymlösung zur Zuckersirupgewinnung werden vorgestellt.

Introduction

Cellulose, a major renewable source of carbohydrate has been widely accepted as a possible substrate for the production of food, fuel or chemicals. Approximately 10^{11} tons of cellulose are synthesized annually and a large percentage of it finds itself as waste. However, most of the wastes occur in combination with hemicellulose and lignin. Advances in the development of technology of bioconversion of cellulose have taken great strides in the last decade [1, 2]. The economics of utilization of cellulose for the production of protein and glucose will be more favorable if steps are taken to generate useful and highly vendable products also from hemicellulose and lignin by-products [3]. A schematic representation of the variety of products which can be obtained by microbial conversion of cellulose is presented in Fig. 1.

Naturally occurring cellulosic wastes are generally recalcitrant to microbial attack. A good pre-treatment of the wastes leads to an increase in the rate of cellulose fermentation and also improves the susceptibility of the substrate to enzymatic hydrolysis [4]. Reduction of particle-size to increase the surface area by physical methods and alteration of the crystalline nature of cellulose to amorphous state by chemical treatment are necessary prerequisites for the development of microbial technology of cellulose utilization. Alkaline pre-treatment has been used in all our experiments on cellulose fermentation. Solka-floc or sugarcane bagasse were used as sources of cellulose. Cellulose was powdered in Wiley Mill and then alkali treated before use. Powdered cellulose was mixed with 1 N sodium hydroxide (50 g cellulose per liter of 1 N NaOH) and heated in an autoclave for 15 min. at 121 C. The swollen fibers were then washed with water to remove the excess alkali.

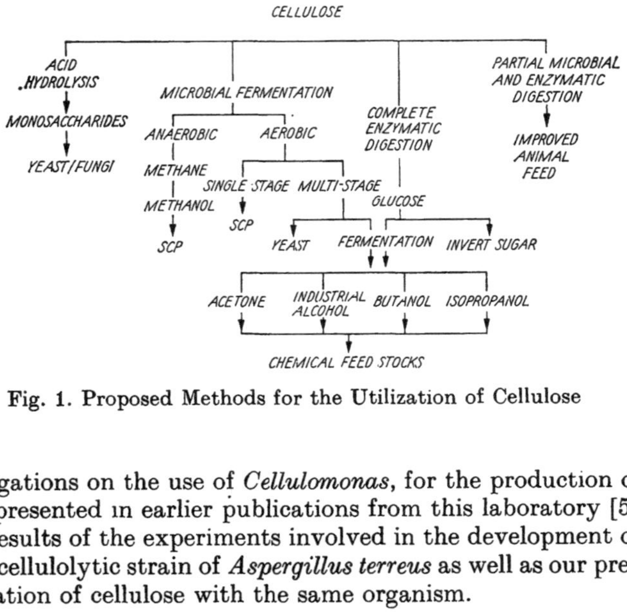

Fig. 1. Proposed Methods for the Utilization of Cellulose

Our investigations on the use of *Cellulomonas*, for the production of Single Cell Protein have been presented in earlier publications from this laboratory [5, 6]. This paper describes the results of the experiments involved in the development of a continuous cultivation of a cellulolytic strain of *Aspergillus terreus* as well as our preliminary attempts at Saccharification of cellulose with the same organism.

Organism: Maintenance and Cultivation

Aspergillus terreus (ATCC 30514) was used in all experiments. The organism was grown on Potato Dextrose Agar (PDA) slants and maintained by transferring every month. Suspensions of conidia for inoculum were obtained from PDA plates containing well sporulated mycelium. The composition of the medium for the growth of *Aspergillus terreus* is presented in Table 1. Each nutrient was proportionately changed with changes in the concentration of the carbon source. Batch cultivation of the organisms was carried out in flasks 50 ml to 1 l. volume and aerated by placing them on gyrotary shakers. The incubation temperature was generally 35 °C. The organism was grown as a continuous culture or as a gradient fed batch culture [5] in a 7 liter stirred tank fermentor (New Brunswick Scientific Co.) with a 4.5—5 l. working volume equipped with foam, level, temperature and *pH* controls. The agitation was varied with the influent carbohydrate concentration (2–10 g/l) and ranged from 400–600 RPM. The aeration was maintained between 1.5 to 2.0 vvm.

Table 1. Composition of the Medium for Growth of *A. terreus*

Nutrient	Concentration mg/liter
Carbon source	
Glucose or cellulose	2000
Ammonium Sulfate	500
Potassium phosphate dibasie	93
Sodium phosphate	31
Magnesium Chloride $6H_2O$	50
Calcium chloride	4
$FeSO_4$ $7H_2O$	1
$ZnSO_4$ $7H_2O$	0.48
$MnCl_2$ $4H_2O$	0.012
$CoCl_2$ $6H_2O$	0.008
$CuSO_4$ $5H_2O$	0.002

Growth of A. terreus on Glucose as Carbon Source — Batch and Continuous

The general growth characteristics of *A. terreus* were determined by growing the organisms as submerged batch cultures. The organism has the ability to grow over a wide range of *pH*'s 3.0–7.0. However all the experiments were carried out at pH 4.0 at which the organism seemed to show a better growth rate. The optimal temperature for growth occurred at 35 °C. with a minimum doubling time of 2 hrs. ($\mu_{max} = 0.346$ h^{-1}). Growth rates at 40 °C and 45 °C were also rapid with minimum doubling times of 2.2 and 2.9 hrs respectively.

Glucose limited continuous culture of *A. terreus* was used to optimize the concentrations of nutrients in the medium. Optimization was conducted by following a modification of the method suggested by MATELES and BALTAT [8]. The organism was extremely sensitive to variations of concentrations of trace metals in the medium. Data from a typical run of continuous cultivation of (glucose limited) *A. terreus* are shown in Fig. 2. Dry

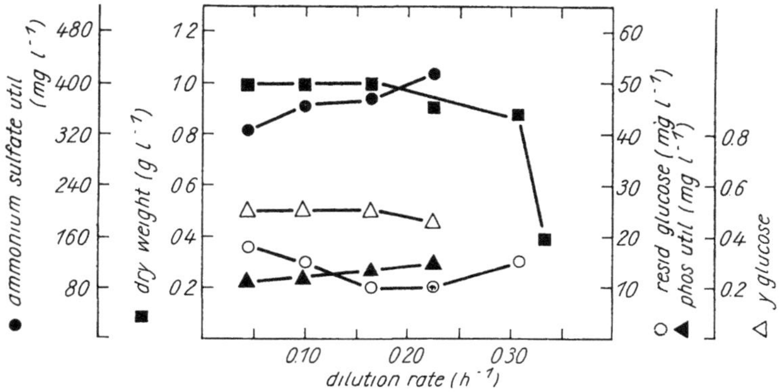

Fig. 2. Growth Characteristics of *Aspergillus terreus* in Glucose-limited continuous Culture

weight values were constant at 1 g/l over dilution rates of 0.047–0.163 hr^{-1}. However, at a specific growth rate of 0.222 hr^{-1} dry weight decreased by 10% although the culture was still glucose-linked. Y glucose ranged from 0.45 to 0.50 over all the examined dilution rates. The maintenance energy co-efficient of glucose for *A. terreus* was calculated to have a value of 15 $mg \cdot g^{-1} h^{-1}$.

Production of Biomass on Cellulose

Table 2 presents results of a few typical experiments on the growth of *A. terreus* on Solka-floc or sugarcane bagasse. There is a significant difference in specific growth rates when the temperature of cultivation is changed from 35–45 °C. Analyses of residual cellulose and protein content in the biomass showed degradation of a large percentage of the cellulose substrate and production of highly proteinaceous biomass in relatively short incubation periods. Results showed that over 88% of the availabe cellulose was consumed in 30 hrs. at 35 °C.

Table 2. Utilization of Cellulose and Production of Protein During Batch Cultivation of *A. terreus*

Expt.	Temp. °C	Sp. Growth Rate h^{-1}	Percent Cellulose Consumed	Percent Protein in Dry Product
1	35	0.057	93.5	37.8
2	35	0.093	88.1	32.9
3	45	0.086	73.3	23.0
4	35	0.084	80.1	29.4

Experiments 1, 2 and 3 were conducted with 10 g/l of alkali-treated Solka-floc while in expt. 4 an equivalent of bagasse was used as substrate.

Experiments were designed to investigate the potentiality of the organism to grow on cellulose in a semi-continuous mode. The organism was grown initially in a modified gradient-fed batch culture. Thereafter every 24 hrs 90% of the culture were harvested and the rest of the culture was retained for the next cycle. Fresh nutrients were added to the fermenter and the growth cycle was repeated. Results of such an experiment are described in Table 3. These experiments indicate that there is no significant change in the specific growth rate of the organism in two successive cycles. There was also no loss in the ability of the organism to utilize cellulose. Nearly 84% of the available cellulose was biodegraded during these experiments. The crude protein content of the dried product ranged from 30–33%.

Table 3. Growth and Production of Protein by *A. terreus* during Semi-Continuous Cultivation of the Organism

Expt.	Temp.	Sp. Growth Rate h^{-1}	% Cellulose Consumed	% Protein in Dry Product
1				
2	35	0.094	84.3	30.0
3	35	0.095	83.7	32.0
4	35	0.091	83.9	32.5

All experiments were run with 10 g/l alkali-treated Solka-floc as substrate

Attempts were made to run a continuous cultivation of *A. terreus* on cellulose. The first and foremost difficulty in running continuous fermentation of cellulose is to pump slurries of cellulose into laboratory fermenters. This was overcome by using low concentration of cellulose suspensions and pumping at short intervals at fast rates to obtain an over-all dilution rate. Experiments were conducted at different dilution rates and temperatures. The population was maintained in a steady state at best for eight residence times at each dilution rate before samples were taken for analyses. Table 4 summarizes the results obtained from a few such experiments. Utilization of cellulase ranged from 74—84% over dilution rates of 0.10—0.14 hr^{-1}. The results indicate that elevated growth temperatures clearly had no adverse effect on the ability of the organism to degrade cellulose. On a commercial scale, substrates may be introduced at 5% level or more. Since we have seen that with the optimized medium (proportionately) increasing all the components of the medium in relation to the carbon source does not change Y_g, it is possible at the present state of the art to obtain a productivity of 3.5—4.0 g/l/h. This makes the process attractive and economically feasible on a commercial scale.

Table 4. Cellulose Utilization During Continuous Cultivation *A. terreus* on Alkali-Treated Solka Floc.

Expt.	Diff. rate h^{-1}	Temp. °C	Influent* Cellulose mg/l	Residual Cellulose mg/l	% Utilization of Cellulose.
1	0.10	35	1500	330	78.0
2	0.11	35	1500	250	83.3
3	0.14	40	1500	270	82.0
4	0.14	45	1500	237	84.2

* A low level of substrate concentration was used in these experiments because of the difficulty of pumping cellulose slurries into 7 liter fermenters continuously.

Saccharification of Cellulose by Aspergillus terreus

Investigations were designed to evaluate the ability of crude cellulases to produce sugar syrups by enzymatic hydrolysis of cellulose. The effect of temperature and *pH* on the saccharification of cellulose is summarized on Table 5. These data show that 45 and

Table 5. Saccharification of Alkali-Treated Solka-Floc by *Aspergillus terreus* cellulase

Hrs Expt			6	12	24
	Temp.	pH	Percent Saccharification		
1	40	4.0	32	61	65
2	45	4.0	39	70	73
3	50	3.0	28	40	43
4	50	4.0	40	65	74
5	50	5.0	33	61	52
6	50	6.0	30	52	53
7	55	4.0	51	61	69
8	60	4.0	38	42	42

Substrate level was 10 g/l

50 °C are optimum temperatures for maximum saccharification at *pH* 4.0. Over 70% of available cellulose was saccharified in 24 hrs. using an initial substrate loading of 10 g/liter. At 55 °C the enzyme seems to be inactivated after 18 hrs. The enzyme concentration present in the medium was estimated to be 0.186 units/ml of culture medium. Maximum production of glucose occurred at pH 5.0 at 50 C, 61% of the cellulose was hydrolyzed in 18 hrs. However, at 24 hrs. bacterial contamination was noticed at *pH* 5.0 and 6.0 since no antibacterial agent was added during the hydrolysis period.

Conclusions

1. HUMPHREY has stated that economic analyses indicate that about a 35% yield is the economic break-even point [9]. In all our experiments, it has been shown that cell yields of 0.38—0.42 g cell/g of cellulose have been consistenly obtained.
2. To date, this is the first report on continuous cultivation of a microorganism with the shortest residence time for the production of biomass from cellulose with 80—85 percent of assimilation of the substrate.
3. Results of our preliminary experiments on the saccharification of cellulose suspensions by crude cellulases were comparable to those obtained by other investigators [10, 11, 12]. Initial scale-up studies have shown that up to 40 g/l of sugar was produced in 24 hrs with over 70% saccharification.

References

[1] "Hydrolysis of Cellulose" ACS Symp. Ser. **181** (1979) Eds. R. D. BROWN and L. JURASEK. Publ. ACS Washington D.C.
[2] Enzymatic Conversion of Cellulosic Materials: Technology and Applications; Biotechnol. Bioeng. Symp. **6** (1976).
[3] M. C. FLICKINGER: Current Biological Research in Conversion of Cellulosic Carbohydrates in Liquid Fuels; Biotechnol. Bioeng. **22** Suppl. 1 (1980) 27.
[4] M. A. MILLETT, Baker, A., and SALTER, L. D.: Physical and Chemical Pre-Treatment for Enhancing Cellulose Saccharification; Biotechnol. and Bioeng. Symp. **6** (1976) 125.
[5] V. R. SRINIVASAN (1975) "Production of Bio-Proteins from Cellulose" in M. BAILEY et al (eds) Symposium on Enzymatic Hydrolysis of Cellulose pp. 393—405 SITRA.
[6] V. R. SRINIVASAN (1979) ,,Production of Single Cell Proteins from Cellulose" in Bioconversion of Organic Residues for Rural Communities pp. 132—138. The United Nations University.
[7] V. R. SRINIVASAN, FLEENOR, M. B., and SUMMERS, R. J.: Gradientfeed Method of Growing high Cell Density Cultures of Cellulomonas in Bunch-Scale Fermenters; Biotechnol. Bioeng. **19** (1977) 153.
[8] R. I. MATELES and BATTAT, E.: Continuous Culture used for Media Optimization; Appl. Microbiol. **28** (1974) 901.
[9] A. E. HUMPHREY, MOREIRA, A., ARRUIGER W., and ZABRISKIE, D.: "Production of Single Cell Protein from Cellulose Wastes"; Biotechnol. Bioeng. Sym. **7** (1977) 45.
[10] D. V. GARCIA MARTINEX, OGAWA T., SHINMYO, A., and ENATSU, T.: Hydrolytic Degradation of Bagasse by Enzymes produced by *Penicillium variabile*; J. Ferment. Technol. **52** (1974) 378.
[11] M. MANDEL'S, HONTZ, L. and NYSTROM, J.: Enzymatic Hydrolysis of Waste Cellulose; Biotechnol. Bioeng. **16** (1974) 1471.
[12] T. G. SHEWALE and SADANA, J. C.: Enzymatic Hydrolysis of Cellulosic Materials by *Sc erotium rotfsii* Culture Filtrates for Sugar Production; Canad. J. Microbiol. **25** (1979) 773.

Acta Biotechnologica 1 (1981) 4, 371–375

Energetic and Carbon Metabolism Regulation in Lysine Producing Brevibacterium Flavum Strains

Y. E. SHVINKA, U. E. VIESTURS, M. P. RUKLISHA, L. A. BABURIN, N. I. GALININA

August Kirchenstein Institute of Microbiology
Latvian SSR Academy of Sciences,
Riga, Latvian SSR, USSR

Paper given at the 2nd Symposium of Socialist Countries on Biotechnology, Leipzig 2.–5. 12. 1980

Summary

Coordinated regulation of carbon and energetic metabolism enzymes is characteristic of lysine producing strains of *Brevibacterium flavum.* ATP is the regulating effector and changes in the stationary ATP-concentration in cells cause certain alternations of enzyme activities in the basic metabolism pathways. The goal of the experiments was the use of the biochemical regulations of methabolism in order to increase the productivity of lysine biosynthesis.
Following results were received:

– Activation of TCA-cycle enzymes is compulsory for intensive lysine biosynthesis.
– The most essential of several parallel electron transport pathways in the ETC of *Br. flavum* is the NADH dependent, cyanid resistant, hydroxamate sensitive oxydation pathway.
– Calculations have shown, that the most economical variant is the synthesis of oxalacetate (precursor of lysine) by PEP carboxylation. Therefore strains with elevated PEP-carboxylase activity synthesize lysine in a more economical way.

Zusammenfassung

Bei Lysin-produzierenden Stämmen von *Brevibacterium flavum* ist eine koordinierte Regulation der Enzyme des Kohlenstoff- und Energiestoffwechsels charakteristisch. Als regulierender Effektor fungiert ATP, so daß sich Veränderungen der stationären ATP-Konzentration auf die Enzymaktivitäten der Hauptstoffwechselwege auswirken. Das Ziel der Untersuchungen bestand in der biochemischen Beeinflussung des Stoffwechsels zur Erhöhung der Produktivität der Lysin-Biosynthese.
Dabei wurden folgende Ergebnisse erhalten:

– Die Aktivierung der TCA-Enzyme ist notwendig verbunden mit einer intensiven Lysin-Biosynthese.
– Der wichtigste von verschiedenen parallelen Elektronentransportwegen in der Elektronentransportkette von *Br. flavum* ist der NADH-abhängige, Cyanid resistente, Hydroxamat-sensitive Oxydationsweg.
– Entsprechenden Berechnungen zufolge ist es energetisch am günstigsten, Oxalacetat (Präkurser für Lysin) durch Carboxylierung von PEP zu erhalten. Deshalb synthetisieren Stämme mit erhöhter PEP-Carboxylase-Aktivität Lysin am effektivsten.

Basic Principles of Metabolism Regulation in *Brevibacterium Flavum*

There are several possibilities for facultative aerobic homoserine-deficient strains *Brevibacterium flavum* to synthesize lysine. The precursor of lysine-oxalacetate is synthesized: 1) from malate, formed in the tricarbonic acid cycle (TCA-cycle); 2) from malate, formed in the glyoxylate cycle (in *Br. flavum* glyoxylate cycle does not function at cultivation on carbohydrates, but is induced by acetate [3]); 3) by phosphoenolpyruvate (PEP) carboxylation. Earlier the material-energetic balance equations for these biosynthesis pathways were formulated and the maximum lysine yield from carbohydrates and acetate — $Yp\left[\frac{\text{g lysine}}{\text{g substrate}}\right]$, was calculated. Yp^{max} values are 0.57, 0.60 and 0.69 for variants 1, 2 and 3 respectively [7].

Simultaneously with lysine synthesis TCA-cycle intermediates are involved in the energetic metabolism of bacteria.

A coordinated regulation of carbon and energetic metabolism enzymes is characteristic of *Br. flavum* [8]. It is supposed, that adenosine triphosphate (ATP) is the regulating effector. Changes in the stationary ATP concentration in cells cause certain alterations of enzyme activities in the basic metabolism pathways (Table 1). Thus, the inhibition of TCA-cycle enzymes causes an increase of the glycolysis enzyme activity and activated glucose transport. The activity of glycolysis increases just as to compensate energetic losses in TCA-cycle preserving the rate of ATP synthesis constant [8]. The coupling

Table 1. Activity of the Basic Metabolic Pathways (in arbitrary units) and Lysine Yield in Bacteria on Various Coupling States of Energetic and Constructive Metabolism

State	Culture characteristics	Carbohydrate transport	Glycolysis	PEP-carboxylase	TCA-cycle	ETC	Alternative oxydases	Yp $\frac{\text{g lysine}}{\text{g substrate}}$
1	Initial culture at the moment of the maximum activity of lysine synthesis (on the early stationary growth phase)	1.5	1.5	0.6	0.8	0.8	0.2	0.40
2	Fully coupled	1.4	1.3	0.7	1.0	1.0	0	0.50
3	Uncoupled phosphorylation in ETC	2.5	2.8	0.9	0.9	0.7	0.4	0.40
4	TCA-cycle enzymes inhibited	3.0	3.5	0.8	0.4	0.2	1.0	0.32
5	PEP-carboxylase inhibited	2.8	3.2	0.5	0.3	0.2	0.8	0.20

Transport activity is determined as $^{14}C_6$ glucose uptake rate. Test-enzymes: for glycolysis — phosphofructokinase (PFK); for TCA-cycle — isocitrate dehydrogenase (ICDH); for electron transport chain — sensitive to 1.10^{-2} M KCN oxygen consumption; for alternative oxydases — resistant to 1.10^{-2} M KCN oxygen consumption. Methods for determination of enzyme and respiratory activity were described previously [1—5].

states of energetic and biosynthetic reactions, shown in Table 1, can be observed after various physiological manipulations: by changing the intensity of aeration, concentration of carbon substrate, after the effect of uncouplers of phosphorylation and electron transport inhibitors (Table 2).

We used the principles of biochemical regulation of metabolism in Brevibacterium flavum in order to increase the productivity of lysine biosynthesis. Several possibilities had been investigated.

Table 2. *Brevibacterium flavum* Key Enzyme Activities after Effect on Cells (A) KCN in Various Concentrations and (B) Superoxide Generating System

Conditions for cell incubation	Glucose transport, rate, imp/min. mg cells	Enzyme activity catal. 10^{-11}/mg protein		
		PFK	PEP-c	iCDH
A. After cell incubation during 20 minutes in the presence of KCN of various concentrations				
Without inhibitor	1430	226	96	86
$+1.10^{-2}$ M KCN	1220	456	43	3
$+1.10^{-3}$ M KCN	1422	446	87	10
$+5.10^{-4}$ M KCN	1840	429	122	84
$+5.10^{-5}$ M KCN	1440	309	115	168
B. After effect of superoxide anionradical (O_2^-) generating system*				
Control	1332	197	80	93
$+O_2^-$ generating system	1242	217	.114	146

* Cells were incubated 2.5 hours in the presence of $2 \cdot 2 \cdot 10^{-6}$ M hypoxantine and 0.2 unit/30 ml xantine oxidase (Olaine)

Activation of TCA-cycle Enzymes

High TCA-cycle enzyme activity is compulsory for intensive lysine synthesis [1, 2, 4]. The activity of TCA-cycle enzymes is highest in state 2 (Table 1); i.e. — under conditions of coupled metabolism. At the moment of the maximum lysine synthesis activity in the early stationary growth phase ($18^{th}-22^{nd}$ hour of cultivation) in *Brevibacterium flavum* the coupling of metabolism is not complete — state 1. An excess of reduced equivalents is synthesized, which are oxidized by the alternative non-phosphorylating pathway. It is possible to activate isocitrate dehydrogenase (TCA-cycle test-enzyme) by means of adding small concentrations of electron transport inhibitor KCN (5.10^{-5} M) into the initial culture (Table 2A). Activation of TCA-cycle enzymes, as compared to the initial culture, can be also provoked in the result of uncoupling of phosphorylation in ETC (state 3, Table 1) or by the effect on cells of the enzymatic system of superoxide radical generation (Table 2B). Activation of isocitrate dehydrogenase in the abovesaid cases causes an increase in the rate of lysine synthesis. Yet the increase of lysine yield (*Yp*) is insignificant. *Yp* does not increase in these cases, because it is difficult to introduce and maintain the culture in coupled state. Under various influences, the culture easily passes into state 3; i.e. with uncoupled phosphorylation in ETC. In this state the rate of glucose transport significantly increases, and the activity of glycolysis is higher. Though the rate of lysine synthesis is high, the lysine yield does not change significantly.

Inhibition of Cyanide-resistant Alternative Oxidation Pathway in ETC

Earlier the simultaneous functioning of several parallel electron transport pathways in *Br. flavum* ETC was established. The most essential one is the NADH-dependent, cyanide-resistant, hydroxamate-sensitive oxidation pathway. Up to 40% of oxygen are consumed in cells by this way, which is not sensitive to 1.10^{-2} M KCN [5]. The resistance of respiration to cyanide was studied in bacteria, grown in chemostat at various dilution rates (fig. 1). KCN-resistant oxygen consumption increases at dilution rates $D > 0.2\ hr^{-1}$. Simultaneously the Yp decreases. It was possible to assume, that a decrease in the KCN-resistant oxygen consumption would lead to an increase of the economy of lysine synthesis.

Cyanide-resistant oxygen consumption in *Br. flavum* is inhibited by hydroxamic acids [5]. The results (Table 3) show, that by adding salicyl hydroxamate (2.10^{-3} M) into

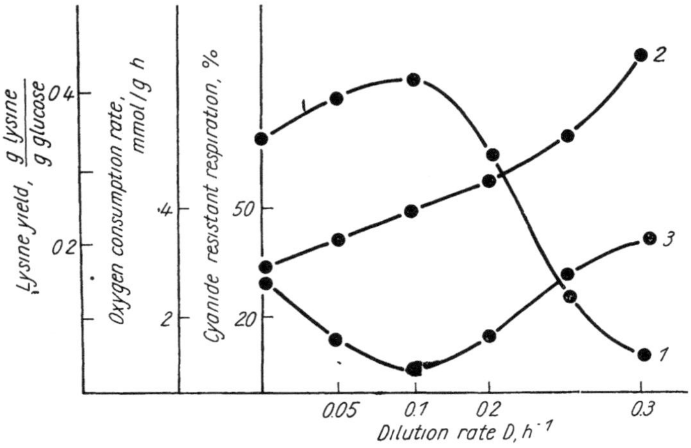

Fig. 1. Lysine yield (1); oxygen consumption rate (2); cyanide-resistant respiration (3) in *Brevibacterium flavum*, cultivated in chemostat at various dilution rates.

Table 3. Effect of Salicyl Hydroxamate* on the Growth, Lysine Biosynthesis, Glucose Transport Rate and Oxygen Consumption Rate (Q_{O_2}) in *Brevibacterium flavum*

Cultivation conditions	Biomass, g/l	Lysine, g/l	Glucose transport rate, imp/min. · mg cells	Q_{O_2}, m mol/g · hr	Yp, g lysine / g substrate
Control	8.7	11.7	1313	3.67	0.40
+ salicyl hydroxamate 2.10^{-3} M	9.3	13.5	1117	2.84	0.47

* Salicyl hydroxamate (2.10^{-3} M) was added on the 6th hour of cultivating bacteria on synthetic medium; cells were cultivated 12 hours in the presence of inhibitor

the cells, grown on synthetic medium, actually cause a higher lysine yield. Besides, except the inhibition of the alternative pathway of oxygen consumption, salycil hydroxamate inhibits glucose transport by 12%.

Activation of PEP Carboxylase

Calculations of material-energetic balance of lysine biosynthesis pathways have shown, that theoretically the most economical is the variant, when oxalacetate is synthesized by phosphoenylpyruvate carboxylation. In this variant Yp^{max} is much greater than in the others — 0.69 [7]. This could serve as a basis for the assumption that strains with an elevated PEP carboxylase activity will synthesize lysine in a more economical way. We tested this principle by selecting the most perspective lysine producers. The indicator was the relation of activity of PEP-c to iCDH at the moment of the maximum lysine synthesis rate (Table 4). In strain RC-115 [6], where this relation is 1.1, lysine yield reaches $Yp = 0.49$.

The discussed examples prove that metabolism regulation of amino acid producing bacteria is a promissing approach for increasing the productivity of lysine biosynthesis.

Table 4. Lysine Yield (Yp), Oxygen Consumption Rate (Q_{O_2}) and Glucose Transport Rate $\left(\frac{dS}{dt \cdot X}\right)$ in *Brevibacterium flavum* Strains, having various Ratios of PEP-carboxylase to Isocitrate Dehydrogenase activities at the Moment of the Maximum Lysinesynthesizing Activity

Strain of bacteria		PEP-c/iCDH	$\frac{dS}{dt \cdot X}$, m mol/ g · hr	Q_{O2}, m mol/ × g · hr	Yp $\frac{\text{g lysine}}{\text{g substrate}}$
Br. flavum	22LD	80/103	1.22	3.79	0.38
Br. flavum	RC-115	87/79	0.98	3.08	0.49

References

[1] RUKLISHA, M. P., VIESTURS, U. E., BEKER, M. E.: Izv. AN Latv. SSR (1976) 3 (344), 90.
[2] RUKLISHA, M. P., VIESTURS, U. E., MEZHINYA, G. R., SELGA, S. E.: Prikladnaya biochimiya i mikrobiologya **2** (1976) 4, 518.
[3] RUKLISHA, M. P., MARAUSKA, D. F., VIESTURS, U. E.: Mikrobiologya **47** (1978) 6, 992.
[4] RUKLISHA, M. P., SHVINKA, Y. E., SAKSE, A. K., VIESTURS, U. E., BEKER, M. E.: Abstracts of the Conference "Microbiological and Enzymatical Synthesis of Amino-acids", Puchshino 1980.
[5] SHVINKA, Y. E., VIESTURS, U. E., TOMA, M. K.: Mikrobiologya **48** (1979) 1, 10.
[6] SHKAGALE, L. B., DALKA, R. B., UDROVSKIS, G. A.: Mikrobiologitsheskaya promyshlennost (1979) 3, 10.
[7] SHVINKA, Y. E., VIESTURS, U. E., RUKLISHA, M. P.: Biotechnol. Bioeng. **22** (1980), 897.
[8] RUKLISHA, M. P., SHVINKA, Y. E., MARAUSKA, D. F., TOMA, M. K., GALININA, N. I.: Abstracts of VIth Intern. Ferment. Symp. and Internat. Symp. on Yeasts, London, Ontario, Canada 1980, 153.

BOOK REVIEW

Gratschowa, J. M.; Gawrilowa, N. N. und Iwanowa, L. A.

Die Technologie von mikrobiellen Eiweißpräparaten, Aminosäuren und Fetten

Verlag Lebensmittelindustrie, Moskau, 1980
448 Seiten, 119 Abbildungen, 55 Tabellen, 32 Empfehlungen auf sowjetische Fachliteratur (in russischer Sprache)

Bei dieser Veröffentlichung handelt es sich um ein für die Hochschulen der UdSSR, die Studenten in der Spezialisierungsrichtung „Technologie mikrobiologischer Produktionen" ausbilden, zugelassenes Lehrbuch.
Ausgehend vom Benutzerkreis dieses Buches ist es sehr übersichtlich aufgebaut und durch eine verständliche Darstellung charakterisiert.
Nach einer kurzen Einleitung über die Entwicklung und die Bedeutung der mikrobiologischen Industrie besonders in der UdSSR wird im ersten Hauptabschnitt eine ausführliche und allgemeine, für das Verständnis der Technologen an den mikrobiellen Prozessen außerordentlich wichtige Einführung in die Morphologie und Physiologie der Mikroorganismen sowie deren Wachstumskinetik und Metabolismus gegeben. Die theoretischen und praktischen mathematischen Beziehungen werden dabei in die Betrachtungen einbezogen.
Den breitesten Raum nimmt verständlicherweise die Beschreibung der Technologien zur Gewinnung von Eiweißstoffen ein. Das leitet sich einmal aus deren volkswirtschaftlicher Bedeutung und auch der möglichen Übertragbarkeit der dargestellten Prozesse auf die weiteren im Lehrbuch betrachteten Stoffgruppen ab. Behandelt werden dabei sowohl die traditionellen als auch die unkonventionellen Rohstoffe, wobei auch solche nicht ausgespart werden, die bisher nur wissenschaftliches Interesse aber noch keine ökonomische Verwertung gefunden haben (als Beispiel dafür seien die Extrakte aus Stein- und Braunkohle genannt). Für die einzelnen Verfahren werden die Besonderheiten bei der Kultivierung der Mikroorganismen und die technologischen Schemata beschrieben, unterteilt nach der mikrobiellen Verwertung der Hydrolysate pflanzlicher Rohstoffe und Sulfitablauge, weiterer Kohlenhydrate, gereinigter n-Paraffine, Erdöldestillate, Erdgas, Methanol, Äthanol, Phenol u. a. Besonders hervorgehoben ist die Beschreibung der Verfahren zur chemischen Hydrolyse; der Stand und die Probleme bei der volkswirtschaftlichen Nutzung des enzymatischen Zelluloseabbaus werden erwähnt.
In zwei weiteren Abschnitten werden die Grundlagen und die technologischen Besonderheiten bei der mikrobiellen Gewinnung von Biolipiden und essentiellen Aminosäuren vorgestellt. Dem schließt sich ein kurzes Kapitel über Fragen des Arbeitsschutzes und des Schutzes der Umwelt bei den mikrobiologischen Produktionen an. Besonders wird auf die Abwässerreinigung eingegangen.
Bei der Beschreibung der verschiedenen Verfahren werden auch Hinweise über eingesetzte Ausrüstungen gegeben, wobei diese für das technische Verständnis der Technologen umfassender sein könnten. Leider werden Fragen der Automatisierung und Automatisierungsmittel fast überhaupt nicht behandelt.
Neben den eingangs genannten Kreis von Studierenden dürfte das Buch auch für technologisch orientierte Mitarbeiter von Interesse sein, die sich einen Überblick über das breite Anwendungsgebiet des Wissenschaftsgebietes Biotechnologie verschaffen möchten.

W. Kauruff

Acta Biotechnologica 1 (1981) 4, 377–386

Spezielle Anwendungsmöglichkeiten mikrobieller Proteasen

P. HEITMANN und D. MEYER

Institut für Technische Mikrobiologie
1017 Berlin, Alt Stralau 62, DDR

Vortrag gehalten anläßlich des 2. Symposiums der sozialistischen Länder über Biotechnologie, Leipzig 2.–5. 12. 1980

Summary

A brief review of the most important properties and fields of application of microbial proteinases are presented. Special possibilities of an industrial utilization of these enzymes are discussed: in the photographic industry for the production and degradation of gelatin as well as for silverless photography, in the food industry for the improvement of the functional properties of proteins as well as for increasing their nutritive value, in organic synthesis for the production of peptides, and in sewage treatment plants for the stabilization of sludge.

Zusammenfassung

Ausgehend von einer Zusammenstellung der wichtigsten Eigenschaften und Einsatzgebiete mikrobieller Proteasen werden spezielle Möglichkeiten der industriellen Nutzung dieser Enzyme behandelt: in der Filmindustrie zur Herstellung und zum Abbau von Gelatine sowie zur Erzeugung silberfreier photographischer Abbildungen, in der Lebensmittelindustrie zur Verbesserung der funktionellen Eigenschaften von Proteinen sowie zur Erhöhung ihres ernährungsphysiologischen Wertes, in der organischen Synthese zur Gewinnung von Peptiden, in Kläranlagen zur Schlammstabilisierung.

Als spezifische und unter milden Bedingungen wirkende Katalysatoren haben Enzyme in den letzten Jahrzehnten eine zunehmende Bedeutung für verschiedene industrielle Umwandlungsprozesse und andere Anwendungszwecke erlangt. Gegenwärtig sind etwa 2000 Enzyme bekannt, aber erst weniger als 20 sind kommerziell tatsächlich bedeutsam. Das Enzym-Weltmarktvolumen dürfte heute in der Größenordnung von einer halben Milliarde DM Umsatzwert liegen [8]. An der Spitze stehen dabei mit etwa 50% die Proteasen (Abb. 1) [20].

Als Quelle für industrielle Enzympräparate, insbesondere für die im Tonnenmaßstab produzierten, kommen aus ökonomischen Gründen in der Hauptsache Mikroorganismen in Betracht. Zudem bieten Mikroorganismen auf Grund der Möglichkeit der Selektion neuer Species und Stämme die Chance der Entdeckung maßgeschneiderter Enzyme für bekannte und neuartige Applikationen, ganz zu schweigen von den sich im Zusammenhang mit der Entwicklung des Geningenieurwesens eröffnenden Möglichkeiten, mit ihrer Hilfe z. B. Humanenzyme für die Enzymsubstitutionstherapie zu gewinnen.

Nach Angaben von K. AUNSTRUP [3] liegt die Gesamtmenge der außerhalb des sozialistischen Wirtschaftgebietes produzierten mikrobiellen Enzyme, bezogen auf reines Enzymprotein, bei etwa 1200 t. Auch in dieser Bezugsskala nehmen die Proteasen mit über 500 t den ersten Platz ein, gefolgt von den stärkeabbauenden Enzymen α-Amylase und Glucoamylase mit etwa je 300 t. Etwa 90% der mikrobiell erzeugten Proteasen werden durch Vertreter der Gattung Bacillus produziert, darüberhinaus sind Streptomyceten sowie die Pilze der Taxa Aspergillus, Mucor und Endothia von Bedeutung. Es handelt sich um extrazelluläre Enzyme, die von den Mikroorganismen in das Nährmedium ausgeschieden werden und sich ohne aufwendigen Zellaufschluß wirtschaftlich relativ günstig gewinnen lassen.

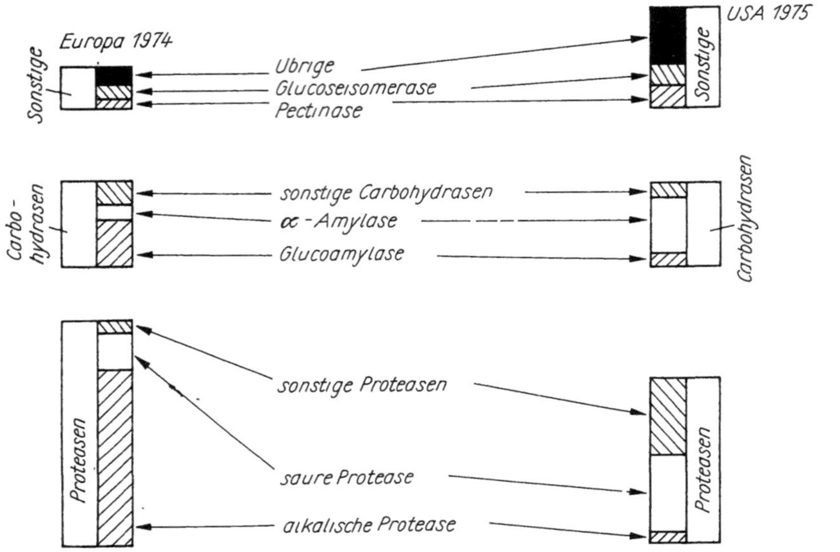

Abb. 1. Anteilmäßiger Umsatzwert verschiedener Enzympräparate in Westeuropa und USA (nach Solomons [20]).

Die Klassifizierung der umfangreichen Gruppe der Proteasen erfolgt nach dem Angriffsort am zu spaltenden Eiweißmolekül (Exopeptidasen, Endopeptidasen) sowie auf der Grundlage des Feinmechanismus der Enzymwirkung. Nach der charakteristischen funktionellen Gruppe im aktiven Zentrum des Enzymmoleküls unterscheidet man vier Proteasetypen: Serin-Proteasen, SH-(Thiol)-Proteasen, saure oder Carboxyl-Proteasen und Metalloproteasen. Für den praktischen Einsatz von Proteasepräparaten ist diese Einteilung jedoch nur von begrenztem Wert. Wichtiger sind in den meisten Fällen die Bedingungen, besonders das *pH*-Milieu, unter denen die betreffende Protease bei hin-

Tabelle 1. Übersicht über verschiedene Proteasetypen

Protease-Typ	pH-Optimum	industriell bedeutsame mikrobielle Produzenten
Serin-Proteasen	schwach bis stark alkalischer Bereich	*Bacillus, Streptomyces; Aspergillus*
Thiol-Proteasen	neutraler bis schwach alkalischer Bereich	*Clostridium; Streptococcus*
Carboxyl-Proteasen	schwach bis stark saurer Bereich	*Aspergillus; Mucor, Endothia*
Metallo-Proteasen	neutraler bis schwach alkalischer Bereich	*Bacillus; Streptomyces; Aspergillus*

reichender Stabilität ihre optimale Wirksamkeit entfaltet. Dabei existieren gewisse Beziehungen zur vorgenannten Einteilung. Serin-Proteasen sind im alkalischen bis neutralen pH-Bereich, Carboxyl-Proteasen dagegen im sauren Milieu am aktivsten, während das *pH*-Optimum der Metallo- und Thiol-Proteasen in der Regel nahe dem Neutralpunkt liegt (Tab. 1). Prinzipiell ist es möglich, für jeden *pH*-Wert innerhalb des Bereichs von etwa 2–12 Proteasen mit ausreichender Stabilität zu finden (Abb. 2). Zunehmendes Interesse finden solche Proteasen, die sich durch eine hohe Thermostabilität auszeichnen, d. h. bis zu Temperaturen um 80–90 °C aktiv sind [18]. Zu berücksichtigen ist auch die Spaltungsspezifität der Proteasen, d. h. die von Fall zu Fall unterschiedlich stark ausgeprägte Fähigkeit zur bevorzugten Spaltung solcher Peptidbindungen, die N- oder/und C-seitig aus bestimmten Aminosäuren (z. B. basische, hydrophobe) aufgebaut sind.

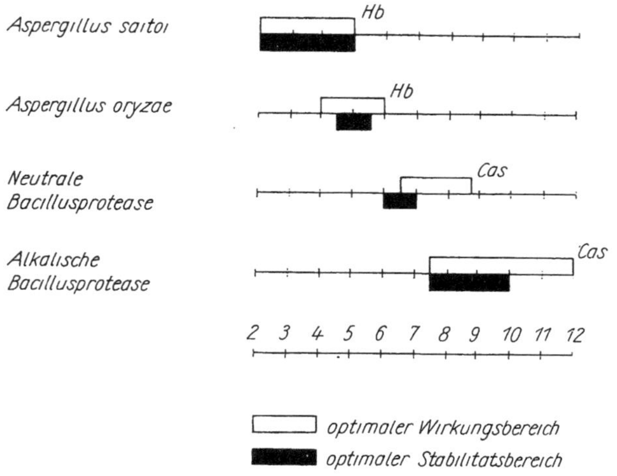

Abb. 2. pH-Aktivitäts- und Stabilitätsoptima verschiedener mikrobieller Proteasen. Als Substrat für die Aktivitätssteigerung diente Hämoglobin (Hb) bzw. Casein (Cas).

Die Tabelle 2 gibt einen Überblick über die wichtigsten Anwendungsgebiete für Proteasen. Länger bekannt ist ihr Einsatz bei der Herstellung und Veredelung von Lebens- und Genußmitteln. Quantitativ an der Spitze steht jedoch heute die Verwendung von alkalischen Bacillus-Proteasen als wirksame Komponenten sog. biologisch aktiver Waschmittel. Der zur Verfügung stehende Raum läßt eine systematische Behandlung der verschiedenen Einsatzgebiete nicht zu. Hier kann auf entsprechende zusammen-

Tabelle 2. Übersicht über wichtige Anwendungsgebiete von Proteasen

Lebensmittelproduktion	Chemische und Leichtindustrie	Sonstige
– Käse	– Wasch- und Reinigungsmittel	– Tierernährung
– Backwaren	– Leder- und Rauchwaren	– Abproduktverwertung
– Fleisch/Fisch	– Gelatine	– Medizin
– Eiweißhydrolysate	– Filme	– Chemische Analytik
– Getränke	– Textilien	
	– Kosmetik	

fassende Darstellungen hingewiesen werden [17, 19]. Die nachstehenden Ausführungen beziehen sich daher im wesentlichen auf einige sich abzeichnende und teilweise schon praktizierte Anwendungsmöglichkeiten für Proteasen, die nach unserer Meinung ein besonderes Interesse beanspruchen.

Als wichtigstes proteasehaltiges Präparat wird in der DDR gegenwärtig im VEB Prowiko Schönebeck aus *Bacillus subtilis* das sogenannte Brauereienzym hergestellt. Auf Grund seines Gehaltes an einer speziellen, Gerstenglucan abbauenden β-Glucanase ermöglicht sein Einsatz in der Brauindustrie die Anwendung einer neuartigen Bierherstellungstechnologie, die durch die Verwendung hoher Anteile ungemälzter Gerste gekennzeichnet ist. Für das Enzympräparat wird ein hoher Proteasegehalt angestrebt, da durch den Abbau schützender Gerstenproteine der Angriff der β-Glucanase erleichtert

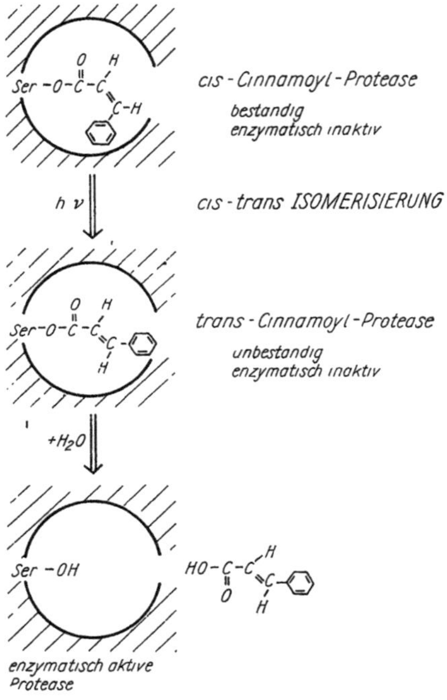

Abb. 3. Lichtinduzierte Aktivierung einer durch Blockierung der aktiven Seryl-Hydroxylgruppe inaktivierten Serin-Protease.

und gleichzeitig der Anteil an assimilierbarem Stickstoff in der Bierwürze erhöht wird. Gegebenenfalls läßt sich dieser Effekt durch den Zusatz spezieller Proteasepräparate noch steigern. Proteolytisch wirksam ist dabei vor allem die neutrale Protease, d. h. eine Metalloprotease, die im Unterschied zur alkalischen Serinprotease durch die in der Gerste enthaltenen spezifischen Proteaseinhibitoren nicht gehemmt wird.

Wachsendes Interesse bringt die Filmindustrie der Nutzung von mikrobiellen Proteasen entgegen. Bereits länger eingeführt ist die Entgelatinierung von Photofilmen, die der Rückgewinnung von Silber und Filmunterlagen aus Alt- und Abfallmaterial dient. Traditionsgemäß werden hierfür Pankreas-Enzyme eingesetzt, jedoch erfordert die weltweit angespannte Situation bei der Zurverfügungstellung entsprechenden Drüsenmaterials eine Substitution der tierischen durch mikrobielle Enzyme.

Mit Hilfe proteolytischer Enzyme lassen sich in der Filmindustrie auch völlig neue Prinzipien der Erzeugung silberfreier photographischer Abbildungen verwirklichen. Eine von Berezin u. a. [4] angegebene Möglichkeit beruht auf der Aktivierung des Enzyms durch Lichteinfluß. Eine photoaktive Schicht, bestehend aus Gelatine, Enzym und einer an das aktive Zentrum des Enzyms kovalent gebundenen Chromophorgruppe, z. B. ein Zimtsäurederivat, wird auf eine Unterlage gebracht. Bedingt durch die Blockierung seines aktiven Zentrums liegt das Enzym zunächst inaktiv vor (Abb. 3). Durch die Belichtung erfährt die blockierende Gruppe eine sterische Umlagerung (cis-trans-Isomerisierung), die sie einem hydrolytischen Angriff zugänglich macht. Das Enzym wird aktiviert und hydrolysiert die Gelatine. Es entsteht ein Relief. Die Fixierung erfolgt durch Inaktivierung des Enzyms. Die verbleibende Gelatine kann gefärbt werden oder dient als Druckmatrize. Bei einer anderen Variante nutzt man die esterspaltende Wirkung von Proteasen und läßt das durch Belichtung aktivierte Enzym nicht auf Gelatine, sondern auf Substrate einwirken, bei deren Hydrolyse zur Farbstoffbildung befähigte Verbindungen freigesetzt werden. Zum Beispiel führt der Indoxylester des N-Acetyl-L-phenylalanins zu Indigo.

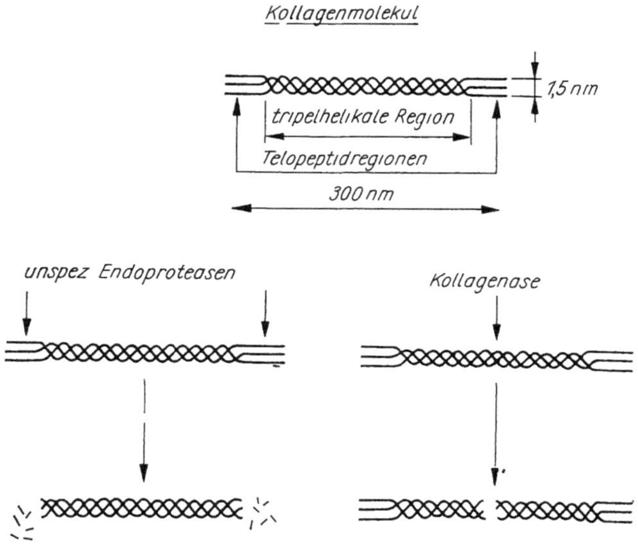

Abb. 4. Aufbau des Kollagen-Monomers (oben) und Angriffsorte verschiedener Proteasen.

Prinzipiell besteht auch die Möglichkeit, in der Photoschicht befindliche aktive Protease unter Vermittlung photosensitiver Farbstoffe, z. B. Bengalrosa, zu inaktivieren und diesen Effekt in analoger Weise wie beim zuerst genannten Verfahren zur Herstellung photographischer Abbildungen auszunutzen [2]. Nachteilig wirkt sich hierbei allerdings die durch die Anwesenheit aktiver Protease bedingte Instabilitat der lichtempfindlichen Schicht aus.

Verschiedentlich sind Versuche beschrieben worden, Proteasen für die Gelatinegewinnung aus kollagenhaltigen Materialien wie Knochen und Tierhäuten einzusetzen [11, 15]. Zwischen den Baueinheiten des Kollagens (Abb. 4) bestehen Vernetzungen kovalenter wie auch nichtkovalenter Art (Abb. 5). Eine wesentliche Rolle bei den kovalenten Vernetzungen spielen die an den beiden Enden der Tripelhelix lokalisierten Telopeptide, die eine nichthelikale Struktur aufweisen. Die proteolytische Spaltung dieser Telopeptide bei weitestgehender Schonung des Helixbereiches stellt offenbar den Primärschritt für die Bildung des monomolekularen Kollagens dar. Die Anwesenheit von Harnstoff oder Guanidiumsalzen wirkt dabei günstig [5]. Anschließend erfolgt durch Hitzeeinwirkung die Auflösung der Helixformation und die Bildung von Gelatine. Vor diesem Schritt muß die Proteaseaktivität vollständig beseitigt werden. Gelingt dies nicht, so kommt es zu einem weiteren proteolytischen Abbau der vergleichsweise wenig strukturierten Einzelketten, was Ausbeuteverluste und Qualitätseinbußen nach sich zieht. Die Proteaseanwendung verfolgt das Ziel, die Nachteile der konventionellen Gelatineherstellung mittels Alkali- oder Säureaufschluß zu vermeiden, bei der Aufschlußzeiten von bis zu mehreren Wochen erforderlich sind bzw. Gelatine ungenügender Qualität entsteht.

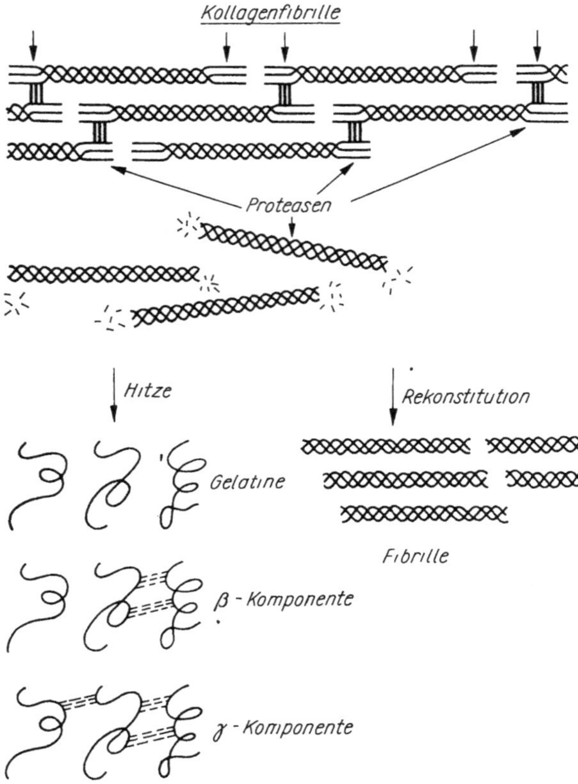

Abb. 5. Vereinfachtes Schema des enzymatischen Kollagenabbaus und der Gelatinebildung.

Die oben erwähnten, durch kontrollierte Auflösung unlöslichen Kollagens erhältlichen löslichen monomolekularen Produkte mit intakter Helixstruktur können zu Fibrillen rekonstruiert werden. Aus den auf diese Weise erhältlichen Dispersionen lassen sich durch Extrudieren mit Hilfe geeigneter Zusätze Garne, Vliese, Folien und Schwämme für lebensmitteltechnische und medizinische Zwecke gewinnen.

Offenkundig sind für die Gelatineherstellung potentiell nur solche Proteasen geeignet, die den Bereich der Tripelhelix unversehrt lassen. Interessanterweise kommen in der Natur auch solche Enzyme vor, die natives Kollagen spezifisch in der helikalen Region angreifen. Verschiedene, insbesondere medizinische Anwendungen werden diskutiert [14]. Ein aus *Clostridium histolyticum* erhältliches Kollagenase-Rohpräparat wird zur Auflösung von nektrotischem Gewebe bei Brand- und anderen Wunden empfohlen.

Zunehmende Anwendung finden Proteasen im Bereich der Nahrungsgüterwirtschaft für die Verbesserung der funktionellen Eigenschaften von Eiweißen [16] wie z. B. ihrer

Löslichkeit [1], aber auch der Viskoelastizität, des Wasserbinde- und Emulgiervermögens. Dabei ist die Aufgabe zu lösen, auf einem kostengünstigen Wege dem Ausgangsprotein die gewünschten funktionellen und organoleptischen Eigenschaften zu verleihen, ohne den ernährungsphysiologischen Wert des Eiweißes wesentlich zu beeinträchtigen. Der partielle proteolytische Abbau erlaubt es, möglichen negativen Auswirkungen der Nahrungsmittelverarbeitung entgegenzuwirken. Auch können auf diese Weise unkonventionelle, d. h. als wenig geeignet angesehene Nahrungsproteinquellen wie z. B. bestimmte Pflanzen- und Einzellereiweiße in einem größerem Umfang als bisher für die menschliche Ernährung erschlossen werden.

In diesem Zusammenhang verdient die bereits aus dem vorigen Jahrhundert bekannte und in den letzten Jahren intensiver untersuchte Plastein-Reaktion Beachtung [6, 7]. Sie umfaßt zwei Stufen: 1. die enzymatische Hydrolyse des Ausgangsproteins und 2. die ebenfalls durch Proteasen (Pepsin, Papain, mikrobielle Enzyme) katalysierte Synthese von Plastein aus den im ersten Teilschritt gebildeten Proteinbruchstücken (Abb. 6). Als

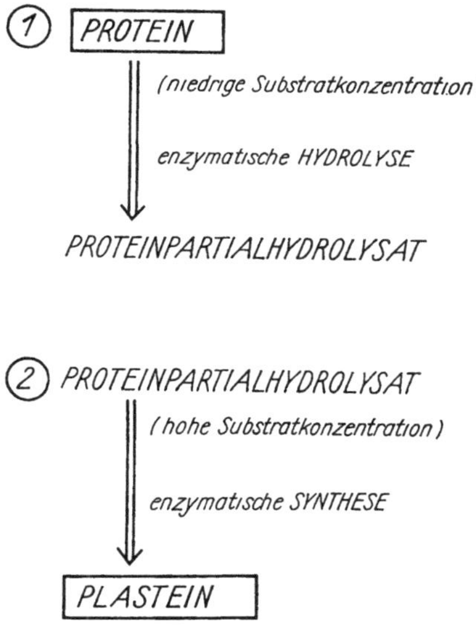

Abb. 6. Plastein-Reaktion

der Plastein-Synthese zugrunde liegende Reaktionsmechanismen werden sowohl eine in Umkehr der Hydrolyse-Reaktion stattfindende Kondensation von Proteinbruchstücken, d. h. eine Neubildung der Peptidbindung, als auch eine Transpeptidierung diskutiert. Nichtkovalenten Bindungen, vor allem hydrophoben Wechselwirkungen wird bei der Plastein-Bildung ebenfalls Bedeutung beigemessen [9]. Beispielsweise wurde vorgeschlagen, die Plastein-Reaktion zur Geschmacksverbesserung von Sojaeiweiß einzusetzen. Unerwünschte Begleitstoffe werden dabei nach dem Proteolyseschritt durch Extraktion entfernt. Bei der Hydrolyse gebildete Bitterpeptide können auch mit Hilfe von Exopeptidasen abgebaut werden.

Eine besondere Bedeutung erhält die Plastein-Synthese durch die Möglichkeit, Aminosäuren in Form ihrer Ester in die Reaktion mit einzubeziehen und auf diese Weise in die entstehenden Produkte zu inkorporieren (Abb. 7). Verwendet man für diesen Zweck Aminosäuren mit hydrophilen Seitenketten, so ist es möglich, die Wasserlöslichkeit beispielsweise von Sojaprotein positiv zu beeinflussen. Durch Inkorporation essentieller Aminosäuren wie Methionin, Threonin, Tryptophan läßt sich der ernährungsphysiologische Wert von Eiweißen steigern [23]. Neuerdings ist es gelungen, den Amino-

säureeinbau in einem Einstufenprozeß zu realisieren und, ausgehend von einem racemischen Gemisch, die L-Komponente des Methioninethylesters zu über 80% in Sojaprotein zu inkorporieren [22].
Weiterhin wurden erfolgreiche Versuche zur Synthese von Peptiden mit Hilfe von Proteasen beschrieben [12] (Abb. 8). Ausgehend von einer N- sowie einer C-geschützten Aminosäure — mit entsprechenden Schutzgruppen versehene kleinere Peptide können ebenfalls in die Reaktion einbezogen werden — erfolgt im wäßrigen Milieu die Knüpfung der Peptidbindung in Ausbeuten bis zu über 80%. Das Reaktionsprodukt fällt dabei

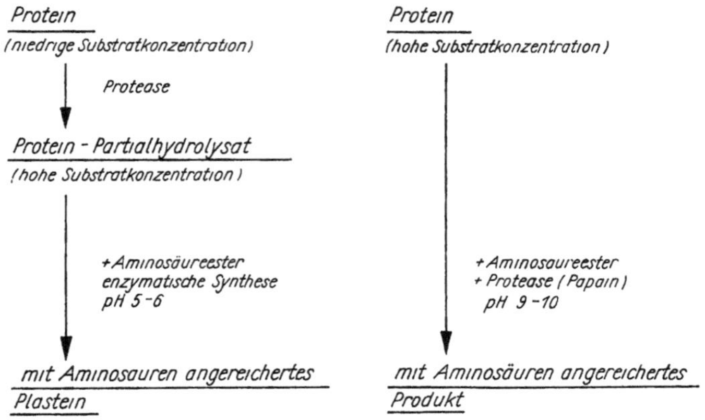

Abb. 7. Aminosäureinkorporation mit Hilfe der Plasteinreaktion im Zweistufen- (links) und Einstufenverfahren.

aus und wird dem Gleichgewicht entzogen. In bestimmten Fällen kann die Peptidbildung durch die Zugabe von organischen Lösungsmitteln begünstigt werden [10]. Vorteilhaft für die Umkehrung der Hydrolyse ist die Verwendung von immobilisiertem Enzym in einem mit Wasser nicht mischbaren organischen Lösungsmittel. So wird bei der Synthese von N^{α}-Acetyl-L-tryptophan-ethyl-ester aus N^{α}-Acetyl-L-tryptophan und Ethanol mit Hilfe von in Chloroform suspendiertem immobilisiertem Chymotrypsin ein praktisch hundertprozentiger Umsatz erreicht [13].
Abschließend sei noch auf das Problem der Beseitigung organischer Abfälle eingegangen, das in hochindustrialisierten Ländern mit ihren großen Städtekonzentrationen und landwirtschaftlichen Intensivviehhaltungen eine immer größere Rolle spielt und für

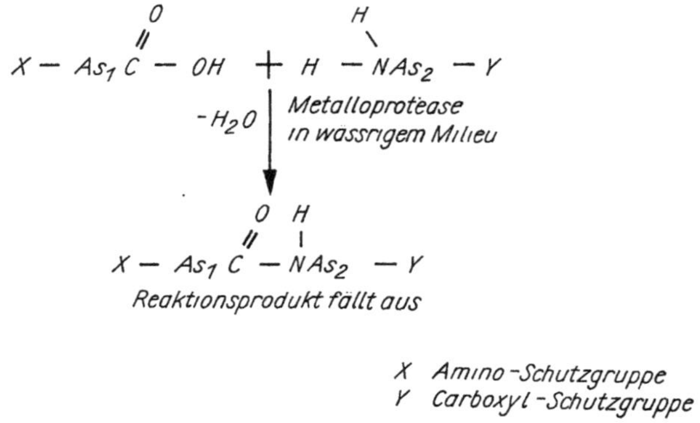

Abb. 8. Peptidsynthese mit Hilfe von Proteasen.

dessen Lösung der Einsatz von Enzymen ebenfalls einen Beitrag liefern kann. Vor allem geht es dabei um eine Beschleunigung der in den verschiedenen Stufen der Abfallaufbereitung ablaufenden Prozesse. Besondere Beachtung verdient in diesem Zusammenhang eine in der DDR untersuchte Möglichkeit der Schlammstabilisierung in Kläranlagen und der Güllebehandlung, bei der sich Proteasen bzw. proteasehaltige Enzymkombinationen als besonders wirkungsvoll erwiesen haben [21]. Auf diese Weise eröffnet sich ein Weg, den zur Stabilisierung von Belebtschlamm erforderlichen Zeitbedarf, der bis zu mehreren Wochen betragen kann, auf einige Stunden zu reduzieren, d. h. die Leistungsfähigkeit der Schlammbehandlungsanlagen auf ein Vielfaches zu steigern.

Literatur

[1] ADLER-NISSEN, J.: Enzymatic Hydrolysis of Proteins for Increased Solubility; J. Agric. Food Chem. **24** (1976) 1090.

[2] AUDRAN, R. G. L., HILAIRE, J.-C., BOURDON, J. A. L. and LESTIENNE, A. F. P.: Photographic Processes and Elements; USA-Patent 3515551, ausgegeben am 2. 6. 1970.

[3] AUNSTRUP, K.: New Developments in the Production and Use of *Bacillus Proteases;* Abhandlungen der Akademie d. Wissenschaften der DDR, im Druck

[4] BEREZIN, I. V., VARFOLOMEEV, S. D., NIKOLSKAYA, I. I., AISINA, R. B., MARTINEK, K., KHLUDOVA, M. S. und KAZANSKAYA, N. F.: Verfahren zur Erzeugung einer silberfreien photographischen Abbildung; DDR-Patent 124688, ausgegeben am 9. 3. 1977.

[5] BERG, A., ECKMAYER, Z., MONSHEIMER, R. und PFLEIDERER, E.: Aufbereitungsverfahren von kollagenhaltigem Rohmaterial; BRD-Offenlegungsschrift 2813075, offengelegt am 11. 10. 1979.

[6] ERIKSEN, S. and FAGERSON, I. S.: The Plastein Reaction and its Application: a Review; J. Food Sci. **41** (1976) 490.

[7] FUJIMAKI, M., ARAI, S. and YAMASHITA, M.: in FEENEY, R. E. and WHITAKER, J. R., (Hag.)/ Food Proteins, Improvement through Chemical and Enzymatic Modification; Adv. Chem. Ser. **160** (1977) 156, Am. Chem. Soc., Washington.

[8] HARTMEIER, W.: Immobilisierte Enzyme für die Lebensmitteltechnologie; Gordian (Hamburg) **77** (1977), 202, 231.

[9] v. HOFSTEN, B. and LALASIDIS, G.: Protease-Catalyzed Formation of Plastein Products and Some of Their Properties; J. Agric. Food Chem. **24** (1976) 460.

[10] HOMANDBERG, G. A., MATTIS, J. A. and LASKOWSKI, M.: Synthesis of Peptide Bonds by Proteinases. Addition of Organic Cosolvents Shifts Peptide Bond Equilibria toward Synthesis; Biochem. **17** (1978) 5220.

[11] HOOREMAN, M. und FORGEOT, M.: Verfahren zur Herstellung von Gelatine; BRD-Offenlegungsschrift 2654093, offengelegt am 2. 6. 1977.

[12] ISOWA, Y., OHMORI, M., ICHIKAWA, T., KURITA, H., SATO, M. and MORI, K.: The Synthesis of Peptides by Means of Proteolytic Enzymes; Bull. Chem. Soc. Jap. **50** (1977) 2762.

[13] KLIBANOV, A. M., SAMOKHIN, G. P., MARTINEK, K. and BEREZIN, I. V.: A New Approach to Preparative Enzymatic Synthesis; Biotechn. Bioeng. **19** (1977) 1351.

[14] MANDL, I. (Hag.): Collagenase; Gordon and Breach, Times Publishers, New York 1972.

[15] PETERSEN, B. R. und YATES, J. R.: Verbesserungen bei oder im Zusammenhang mit der Extraktion von Gelatine; BRD-Offenlegungsschrift 2629594, offengelegt am 27. 1. 1977.

[16] RICHARDSON, T. in FEENEY, R. E. and WHITAKER, J. R. (Hag.)/Food Proteins, Improvement through Chemical and Enzymatic Modification; Adv. Chem. Ser. **160** (1977) 185 Am. Chem. Soc., Washington.

[17] RUTTLOFF, H., HUBER, J., ZICKLER, F. und MANGOLD, K.-H.: Industrielle Enzyme; VEB Fachbuchverlag, Leipzig 1978, 1. Aufl.

[18] RUTTLOFF, H., KLINGENBERG, P., BEHNKE, U., LEUCHTENBERGER, A. und TÄUFEL, A.: Gewinnung und Charakteristik von Proteasen aus *Thermoactinomyces vulgaris*, I. Proteasen aus thermophilen Mikroorganismen (Literaturüberblick); Z. allg. Mikrobiol. **18** (1978) 437.

[19] Schreiber, W., Schindler, J. und Gloxhuber, C.: Herstellung und Verwendung mikrobieller Proteasen; Chemiker-Ztg. **98** (1974) 539.
[20] Solomons, G. L.: Microbial Production of Enzymes; in Sharon, Z. B. N. (Hag.)/Biotechnological Applications of Proteins and Enzymes, Academic Press, New York 1977.
[21] Trommler, K., Riedel, W. und Friedrich, E.: Verfahren zum Abbau von Belebtschlamm und Gülle; DDR-Patent 140244, ausgegeben am 20. 2. 1980.
[22] Yamashita, M., Arai, S., Amano, Y. and Fujimaki, M.: A Novel One-step Process for Enzymatic Incorporation of Amino Acids into Proteins: Application to Soy Protein and Flour for Enhancing Their Methionine Levels; Agric. Biol. Chem. **43** (1979) 1065.
[22] Yamashita, M., Arai, S. and Fujimaki, M.: Plastein Reaction for Food Protein Improvement; J. Agric. Food Chem. **24** (1976) 1100.

Acta Biotechnologica 1 (1981) 4, 387–390

SHORT COMMUNICATIONS

Enzym-Muster Methanol assimilierender Hefen in Abhängigkeit vom Kohlenstoff-Substrat

R. MÜLLER, K. H. HOFMANN und W. BABEL

Akademie der Wissenschaften der DDR
Institut für technische Chemie Leipzig
7050 Leipzig, Permoserstr. 15

Poster anläßlich des 2. Symposiums der sozialistischen Länder über Biotechnologie, Leipzig 2.–5. 12. 1980

Summary

The influence of the carbon substrates glucose and methanol on enzyme activities of the yeast *Candida methylica* was investigated by substrate-shift experiments in discontinuous cultivation. Growth on glucose results in a repression of the enzymes necessary for methanol utilization. After depletion of glucose or addition of methanol these enzymes are derepressed or induced, respectively. Changes in the activity of enzymes of the intermediary metabolism turn out differently both in intensity and direction as a result of substrate-shifts. The results suggest a basic change in the metabolism by growing *Candida methylica* in glucose or methanol.

Zusammenfassung

Der Einfluß der Kohlenstoffstubstrate Glucose und Methanol auf Enzymaktivitäten der Hefe *Candida methylica* wurde mittels Substrat-"shift" – Experimenten in diskontinuierlicher Kultur untersucht. Glucose führt zur Repression der für das Wachstum auf Methanol notwendigen Enzyme, bei Glucoseverbrauch bzw. Methanolzugabe kommt es zur teilweisen Derepression bzw. Induktion. Die Änderungen in der Aktivität von Enzymen des Intermediärstoffwechsels verlaufen bei den Substrat-"shifts" unterschiedlich in Intensität und Richtung. Insgesamt deuten die Ergebnisse auf eine grundlegende Umstellung des Stoffwechsels bei Wachstum auf Glucose bzw. Methanol hin.

Der Einfluß der Kohlenstoff-Substrate Glucose und Methanol auf die Aktivität von Enzymen der Methanol-Dissimilation (Methanol-Oxidase, Katalase, Formiat-Dehydrogenase) und der zyklischen Regenerierungssequenz des C_2-Donors Xylulose-5-phosphat für die Formaldehyd-Fixierung (Dihydroxyaceton-Kinase, Fructose-1.6-bisphosphatase), des Intermediärstoffwechsels ($NADP^+$-spezifische 6-Phosphogluconat-Dehydrogenase, Citratsynthase, $NADP^+$-spezifische Isocitrat-Dehydrogenase, NAD^+-spezifische Malat-Dehydrogenase) sowie der NAD^+-spezifischen Glycerin-Dehydrogenase in der Hefe *Candida methylica* wurden mit Hilfe von Substrat-"shift"-Experimenten in diskontinuierlicher Fermentation untersucht.
Bei Kultivierung auf Glucose wuchs *Candida methylica* mit einer Verdopplungszeit von 2,9 Stunden. Nach Glucoseverbrauch zugesetztes Methanol führte erst nach einer relativ

langen Verzögerungsphase von 10 bis 12 Stunden zu erneutem Wachstum der Kultur (Verdoppelungszeit = 8,2 Stunden), obwohl bereits nach fünf Stunden eine Methanolabnahme im Medium gemessen werden konnte (Abb. 1 A).
Bei Wachstum auf Glucose waren die Enzyme der Methanol-Dissimilation und zyklischen Regenerierung nicht oder nur in geringer Aktivität nachweisbar (Abb. 1 B). Verbrauch der Glucose führte zur Derepression [vgl. 2, 4] bzw. Methanol-Zusatz zur Induktion dieser Enzyme [5].

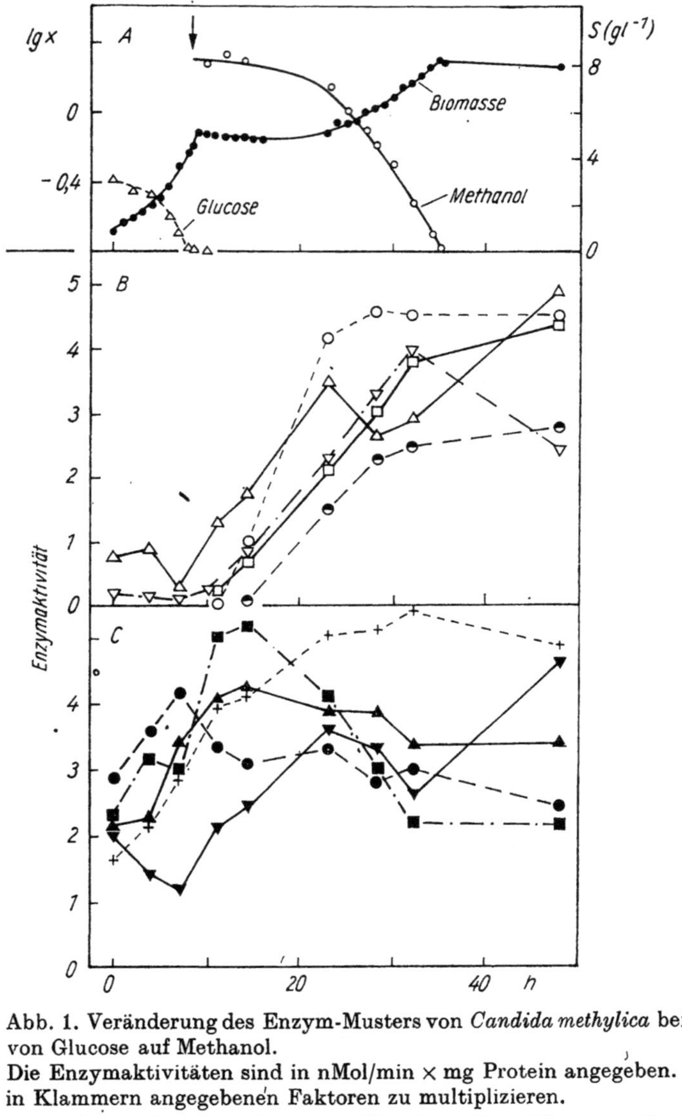

Abb. 1. Veränderung des Enzym-Musters von *Candida methylica* bei Substrat-"shift" von Glucose auf Methanol.
Die Enzymaktivitäten sind in nMol/min × mg Protein angegeben. Sie sind mit den in Klammern angegebenen Faktoren zu multiplizieren.
○ — Methanoloxidase (25), ▽ — Katalase (2 × 10^5), □ — Formiat-Dehydrogenase (50), △ — Fructose-1.6-bisphosphatase (25), ◓ — Dihydroxyaceton-Kinase (100), ● — 6-Phosphogluconat-Dehydrogenase (50), ▼ — Glycerin-Dehydrogenase (25), ■ — Citratsynthase (250), ▲ — Isocitrat-Dehydrogenase (50), + — Malat-Dehydrogenase (3 × 10^3)

Auch die Aktivitäten der Glycerin-Dehydrogenase und der Malat-Dehydrogenase [4] nahmen nach Übergang von Glucose zu Methanol auf das Mehrfache zu, während die Aktivität der Citratsynthase stark abfiel (Abb. 1C).
Nach Auszehrung des Methanols wurden innerhalb von 13 Stunden nur die Katalase und die Malat-Dehydrogenase partiell inaktiviert. Die Aktivitäten der übrigen Enzyme blieben konstant (1B, C).
Zusatz von Glucose zu einer auf Methanol wachsenden Kultur führte nach einer Verzögerungsphase von ca. 3 Stunden zu der erwarteten Steigerung der Wachstumsge-

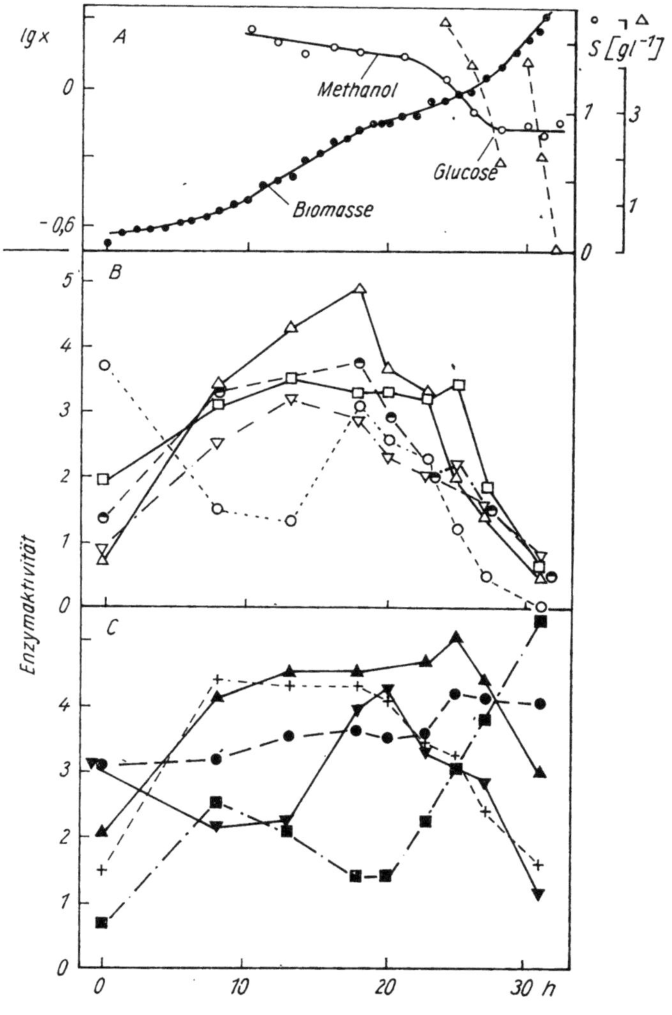

Abb. 2. Veränderung des Enzym-Musters von *Candida methylica* nach Zugabe von Glucose zu einer auf Methanol wachsenden Kultur. Die Angaben der Aktivitäten erfolgten wie bei Abb. 1.

○ – Methanoloxidase (200), ▽ – Katalase (5×10^4), □ – Formiat-Dehydrogenase (50), △ – Fructose-1.6-bisphosphatase (100), ◓ – Dihydroxyaceton-Kinase (100), • – 6-Phosphogluconat-Dehydrogenase (50), ▼ – Glycerin-Dehydrogenase (20), ■ – Citratsynthase (100), ▲ – Isocitrat-Dehydrogenase (25), + – Malat-Dehydrogenase (5×10^3)

schwindigkeit und vorübergehend zu einer starken Erhöhung des Methanolverbrauches (Abb. 2A).

Der Glucosezusatz verursachte bei allen gemessenen Enzymen der Methanol-Assimilation und zyklischen Regenerierung sowie bei der Glycerin-Dehydrogenase und Malat-Dehydrogenase eine starke Abnahme der spezifischen Aktivitäten (Abb. 2B, C). Da im Falle der Methanol-Oxidase, Katalase, Formiat-Dehydrogenase und Dihydroxyaceton-Kinease auch die Gesamtaktivität im Fermentor stark abnahm (nicht dargestellt), muß der Aktivitätsabfall dieser Enzyme zusätzlich zu einer Katabolit-Repression durch Katabolit-Inaktivierung erklärt werden [3]. Die Aktivität der Citratsynthase nahm nach Glucosezusatz erwartungsgemäß zu (Abb. 2C), während die Isocitrat-Dehydrogenase und 6-Phosphogluconat-Dehydrogenase keine deutliche Abhängigkeit von der C-Quelle zeigten (Abb. 1C, 2C).

Eine grundlegende Umstellung des Stoffwechsels von *Candida methylica* ist die Folge eines Wechsels zwischen den Kohlenstoff-Substraten Glucose und Methanol. Davon werden nicht nur die ausschließlich für die Verwertung von Methanol als Kohlenstoff- und Energiequelle notwendigen Sequenzen, sondern auch Enzyme des Intermediärstoffwechsels betroffen. Graduelle Unterschiede im Aktivitätsniveau von Enzymen des Zwischenstoffwechsels bei Wachstum auf Glucose oder Methanol können mit einer mehr oder weniger starken Expression der betreffenden Gene erklärt werden. Darüber hinaus ist die Synthese verschiedener — „heterotropher" oder „methylotropher" — Isoenzyme wahrscheinlich, wie es im Falle der Citratsynthase für fakultativ methylotrophe Bakterien bereits gezeigt wurde [1].

Literatur

[1] Babel, W. und Müller-Kraft, G.: Regulation der Citrat-Synthase bei methylotropher und chemoorganoheterotropher Ernährung von Bakterien; Z. Allg. Mikrobiol. **19** (1979) 687.

[2] Eggeling, L. and Sahm, H.: Derepression and partial Insensitivity to Carbon Catabolite Repression of the Methanol Dissimilating Enzymes in *Hansenula polymorpha;* European J. Appl. Microbiol. Biotechnol. **5** (1978) 197.

[3] Holzer, H.: Catabolite Inactivation in Yeast; Trends Biochem. Sci. **1** (1976) 178.

[4] Schlanderer, G., Held, W., Reimann, J. and Dellweg, H.: Metabolic Changes during Substrate Shiftes in continuous Cultures of the Yeast *Candida boidinii* Variant 60; European J. Appl. Microbiol. Biotechnol. **6** (1978) 133.

[5] Yashuhara, S., Kawamoto, S., Tanaka, A., Osumi, M. and Fukui, S.: Induction of Catalase Activity in a Methanol utilizing Yeast, *Kloeckera* sp. No. 220; Agr. Biol. Chem. **40** (1976) 1771.

Acta Biotechnologica 1 (1981) 4, 391–392

Rezension

2. Symposium der sozialistischen Länder über Biotechnologie vom 2.–5. Dezember 1981 in Leipzig

Das Symposium wurde vom Institut für technische Chemie der AdW der DDR und der Gesellschaft für Allgemeine und Technische Mikrobiologie in der Biologischen Gesellschaft der DDR organisiert und durchgeführt.
Dieses 2. Symposium setzte eine im Jahre 1978 in Riga (UdSSR) begonnene Reihe über Biotechnologie fort.
An dem Symposium nahmen 421 Wissenschaftler aus 12 Ländern teil.
Es verfolgte das Ziel, den Stand und die Perspektiven der Biotechnologie zu erfassen, einen breiten Erfahrungsaustausch der Teilnehmer aus den sozialistischen Ländern zu ermöglichen und die Zusammenarbeit auf dem Gebiet der Biotechnologie innerhalb des RGW weiter zu verstärken.
Auf dem Symposium wurden 6 Plenarvorträge und 45 Vorträge gehalten, 2 Rundtischdiskussionen zu Fragen der Ausbildung von Studenten und der Risikofaktoren der biotechnischen Prozesse durchgeführt sowie 183 Poster demonstriert.
Die Biotechnologie wird als selbständiges Wissenschaftsgebiet allgemein anerkannt und als wissenschaftliche Lehre von den industriellen mikrobiologischen und biochemischen Prozessen sowie solchen mit bioanalogen Strukturen verstanden (M. Ringpfeil).
Sie erlangt zunehmend an Bedeutung für die Gesunderhaltung und Ernährung des Menschen, für die Erschließung neuer Rohstoffe und die Gestaltung der Umwelt, wie übereinstimmend die Eröffnungsredner H. Klare (AdW der DDR), H. Lyr (Biologische Gesellschaft der DDR) und K. Löffler (Ministerium für Chemische Industrie) betonten.
Die Plenarvortragenden vermittelten ausgezeichnete Übersichten über den Stand ausgewählter Teilgebiete der Biotechnologie.
M. Ringpfeil (DDR) charakterisierte das sich neu herausgebildete Wissenschaftsgebiet der Biotechnologie und stellte an instruktiven Beispielen dar, wie sich die Biotechnik der Einzeller neben der Landwirtschaft und dem Gesundheitswesen zum dritten großen Anwendungsgebiet der Biowissenschaften entwickelt hat.
Neue Ergebnisse auf den Gebieten der Biogeologie (M. W. Iwanow – UdSSR), anaeroben Prozesse (M. E. Beker – UdSSR), Nutzung landwirtschaftlicher Biomassen (J. Hollo – VR Ungarn) und energetischen Bilanzierung mikrobiologischer Prozesse (P. Krumhanzl – CSSR) wurden vorgestellt.
Neben den Massenprozessen werden von der Biotechnologie auch zunehmend Anforderungen an Spezialprodukte für die Medizin erwartet (F. Jung – DDR).
Das Symposium umfaßte 6 spezielle Sektionen mit den Themenkreisen:

Sektion 1: Grundlagen mikrobiologischer Prozesse
Sektion 2: Grundlagen biochemischer Prozesse
Sektion 3: Automatisierung, Meßtechnik, Optimierung
Sektion 4: Reaktorgestaltung
Sektion 5: Trennung, Reinigung, Isolation von Produkten
Sektion 6: Applikation von biotechnisch erzeugten Produkten

Das Hauptinteresse der Vortragenden galt Problemen der Energetik und Kinetik und der Optimierung mikrobiologischer Reaktionen. Die genetische Beeinflussung mikrobieller Leistungen entwickelt sich immer mehr als Grundlage für effektivere Möglichkeiten von Biosynthesen. Innerhalb der Produktsynthesen galt das Hauptaugenmerk den Enzym-, Ethanol- und Lysinproduktionen.

Als Massenrohstoffe für mikrobiologische Zellsubstanzsynthesen spielen gegenwärtig Methanol und cellulosehaltige Pflanzenbiomassen die bedeutendste Rolle.
Beeindruckend waren die vorgestellten Leistungskennziffern der entwickelten Umlauffermentoren (DDR) mit sehr hohen Produktivitäten bzw. Sauerstofftransferraten bis 100 kg O_2/t · h.
Zur Trennung biologischer Stoffe und Zellen liegt der Forschungsschwerpunkt auf den energiesparenden mechanischen Methoden und Prinzipien.
Mikrobielle Eiweißfuttermittel, gezüchtet auf chemischen Grundstoffen (Methanol, Kohlenwasserstoffe) zeigten im Ergebnis von Applikationsuntersuchungen ihre hohe Bedeutung für die Tierernährung und ihre vergleichbare Qualität mit anderen herkömmlichen Proteinträgern.
Das 3. Symposium der sozialistischen Länder über Biotechnologie wird im Frühjahr 1981 in der CSSR durchgeführt.

H. J. Schneider; D. Meyer

The Eucaryotic Ribosome

Herausgegeben von Heinz Bielka, mit Beiträgen von einem Autorenkollektiv

— in englischer Sprache —

1981. Etwa 320 Seiten — 36 Abbildungen — 26 Tabellen — gr. 8°
Leinen etwa 40,— M
Bestell-Nr. 762 610 7 (6613)

Ausgehend von einer kurzen historischen Darstellung des Forschungsgebietes werden Vorkommen der Ribosomen in der Zelle (Anzahl, Verteilung, Organisation) und einige wesentliche Eigenschaften im Überblick dargestellt. Im speziellen werden behandelt: Morphologische Eigenschaften der Ribosomen; die makromolekularen Bausteine der Ribosomen; die Organisation, Lokalisation und Funktion der ribosomalen Bausteine im Ribosom; Wechselwirkungen mit Membranen und deren biochemische Grundlagen und Bedeutung; der Gesamtprozeß der Proteinbiosynthese unter Berücksichtigung der am Translationsprozeß beteiligten nichtribosomalen Komponenten.

Bestellungen durch eine Buchhandlung erbeten

THEORIA CUM PRAXI

AKADEMIE-VERLAG

DDR-1086 Berlin, Leipziger Str. 3—4

Contents

Short Communications

Book Reviews

Bei Fragen zur Produktsicherheit wenden Sie sich bitte an:
If you have any questions regarding product safety,
please contact:

Walter de Gruyter GmbH
Genthiner Straße 13
10785 Berlin
productsafety@degruyterbrill.com